# HUMANS 3.0

ALSO BY PETER NOWAK

*Sex, Bombs, and Burgers: How War, Pornography, and Fast Food Have Shaped Modern Technology*

# HUMANS 3.0

*The Upgrading of the Species*

## PETER NOWAK

Guilford, Connecticut

An imprint of Rowman & Littlefield

Distributed by NATIONAL BOOK NETWORK

British Library Cataloguing-in-Publication Information available

Library of Congress Cataloging-in-Publication Data

Nowak, Peter, 1974-
  Humans 3.0 : the upgrading of the species / Peter Nowak.
    pages cm
  Includes index.
  ISBN 978-0-7627-8700-5 (alk. paper)
  1. Technology—Social aspects. 2. Technological innovations—Social aspects. 3. Technology and civilization. 4. Social evolution. I. Title.
  T14.5.N69 2015
  303.48'3—dc23
                                  2014037165

*For my mom*

*"Changes aren't permanent, but change is."*
— *"Tom Sawyer," Rush*

# Contents

# 1

# Evolution: Of Rice and Men

*Boast not thyself of tomorrow for thou knowest not what a day
may bring forth.*

—Proverbs 27:1

Ask an anthropologist what we are as a species, and you might get a response explaining the main branches or epochs of human evolution. Biological developments or environmental changes caused the differences in each step from Australopithecines to Neanderthals to the *Homo* genus. Our brows became less sloped because our brains got bigger, or we stood more upright so that we could run faster, and so on. These little advances decided which of the various human subspecies ultimately won out, although scientists still disagree on exactly how or why this happened. One line of thinking, for example, believes that *Homo sapiens* outlasted Neanderthals because they were better able to adapt to the Ice Age. Whatever the case, we are today—and have been for the past two hundred thousand years or so—the last surviving subspecies of humans: *Homo sapiens sapiens*.

But that doesn't really answer the question. The definition of what we are can't be limited to simple biological evolution because that branch of science doesn't exclusively explain what it is to be human. As any student of psychology knows, people are products of both their biology and environment, which involves their upbringing and surroundings. In this classic nature-versus-nurture dichotomy, the long-debated question is which of the two exerts more influence in determining who we are.

In the early days of human evolution—hundreds of thousands of years ago—it was an even combination of the two. With different subspecies competing for limited resources, our ancestors had to adapt to their environments quickly, hence the bigger brains, better posture, and other physiological changes that took place over a relatively short time period. But, as *Homo sapiens* won out and the others died off, that initially speedy biological evolution slowed down, eventually settling into a snail's pace.

Business cycles behave in the same way. In a given industry, companies compete for market share with a certain product or service. When one emerges the winner and ultimately becomes a monopoly, innovation slows and sometimes stops entirely. *Homo sapiens* are like Microsoft, except they monopolized evolution rather than the home computer software market. If an antitrust regulator back then had quashed that evolutionary monopoly, innovation might have continued at the same pace, meaning we all would have wings and psychic powers by now. But that didn't happen, which is why we're not *Homo superior* or the superpowered mutants found in X-Men comics. Human biological developments in the modern age have been few and slow going, the most notable—though still relatively minor—being further height increases or less body hair. Perhaps in the near future, *Homo superior* will arise: Hairless Man, a superhero who can swim hard, fast, and far when naked.

Today and into the future, it's the second factor—the environment—that's becoming much more important to continued human evolution. This is because technology, the main tool by which people affect their environments, is moving at the opposite pace of biology. Rather than slowing to a trickle, technology's rate of advancement is skyrocketing.

In 1965, chemist and businessman Gordon Moore observed that the number of transistors on an integrated circuit had doubled every year since it was invented in 1958. He estimated the trend would continue for at least another decade. He was right in his basic evaluation but wildly wrong, in a good way, about the time frame. His observation became Moore's Law and has expanded into a golden rule that governs pretty much all of technology. Generally speaking, it posits that the performance and power of just about all forms of technology double every two years or so, yet prices stay the same or decline. This is largely why everything from computers to phones gets more powerful, cheaper, or both every year. It's also why that flat-panel TV you bought for thousands of dollars a few years ago is worth practically nothing today. Moore later cofounded Silicon Valley stalwart Intel, which has kept his theory alive and moving.

Moore's Law has fueled many of the dramatic technological advancements of the past half-century, from the moon landing to the Internet. Just about every digital electronic component benefits from fast and steady improvements in price and performance. New technologies have come into being, and existing ones have gotten better. The real magic, meanwhile, happens when those various components combine in new and interesting ways.

Let's consider a host of random, disparate technologies and gadgets: say, a computer processor, GPS, battery, microphone, camera, speaker, memory storage, graphics chip, wireless radio, and touch screen. Each of these items was derived from decades of research and development, and each became better, smaller, and cheaper over that time. For example, early microphones—big, bulky metallic things—cost a small fortune. Today, they're pin-sized and can be had for pennies. Computer processors and their constituent circuits used to take up entire rooms, whereas now you need a microscope to see them.

What happens when you combine all of those ingredients? A smartphone, one device that really consists of many technologies. A typical smartphone contains thousands of singular innovations ruled by Moore's Law that collectively combine into something far more powerful. They stack on one another, each new invention making possible a whole host of newer inventions. As the stack advances and builds, the rate at which the innovations accrue also increases. The pieces might move slowly at first, but, once the numbers start to build, they multiply rapidly. (Remember the graph of y = x2 from algebra class? That's how the phenomenon looks.)

The best way to understand Moore's Law is through the old story about the invention of chess. A clever inventor takes his game to the emperor of the land, who is so impressed that he tells the man to name his reward. The inventor asks for a quantity of rice as follows: one grain for the first square on the board, two for the second, four for the third, eight for the fourth, and so on, with the amount doubling for each subsequent square. Emperors aren't necessarily math wizards, so the ruler laughs at the reward as small but grants it. The inventor continues, and, after completing the first half of the chessboard, his reward still isn't very impressive. As futurist Ray Kurzweil puts it in *The Age of Spiritual Machines*,

*After thirty-two squares, the emperor had given the inventor about four billion grains of rice. That's a reasonable quantity—about one large field's worth—and the emperor did start to take notice. But the emperor could still remain the emperor. And the inventor could still retain his head. It was as they headed into the second half of the chessboard that at least one of them got into trouble.[1]*

At the end, the inventor ended up with eighteen quintillion grains of rice or enough to get him beheaded in some versions of the story. While apocryphal, the story does illustrate nicely how exponential growth and Moore's Law work. Technologically speaking, the smartphone probably lies somewhere in the middle of that chessboard, which means the really fun stuff is yet to come.

Another more recent, more real example of how exponential growth works is the Human Genome Project. The effort to map the entire human DNA sequence began in 1990 with the expectation that it would take fifteen years. A few years into it, however, scientists fretted. Some believed they'd never finish since they had only completed a percentage of the product in the low single digits. What they didn't count on were rapid advancements in the technologies they were using, such as supercomputers, as well as more people joining the effort with additional resources. Those forces eventually resulted in a massive acceleration of what they initially considered meager progress. Before they knew it, their progress doubled and then doubled again. Their concerns were unwarranted because 1 percent lies only seven doublings from 100 percent. The Human Genome Project crossed the finish line in 2003, two years early.

This exponential growth is having and will continue to have profound effects on the rate of technological change and therefore the environment and humanity. While human biological evolution has largely diminished, the human environment has become increasingly technology-driven. But the lines between biology and the environment are starting to blur.

## THE NEED FOR SPEED

A case in point lies on California's Pacific Coast Highway, where computer scientist and futurist John Seely Brown is driving his brand new Porsche 911S. Like many warm-blooded males, he loves fast and flashy cars. But as a technophile, he's as impressed with the onboard systems as the ride itself. The dashboard display gives him a constant readout of his tire pressure. A rear spoiler emerges automatically at high speeds to put downward pressure on the car to reduce drag and increase fuel efficiency. The integrated GPS displays real-time traffic data as well as the current speed limit on every inch of the road. The smart suspension system practically steers itself through the winding turns.

Seely Brown feels at home in what he calls the Porsche's beautiful harmony. "The sense of being one with the machine and the road—an amplified, extended mind—the ultimate feeling of power and control, all interfaces . . . have phenomenologically disappeared," he says. "I am one with the world."[2] It's the ideal piece of technology: a four-wheeled computation platform with all of its processes hidden behind a sleek facade. When he's behind the wheel, a feeling of Zen overcomes him. The Porsche is more than a car; it's like a second skin that connects its human symbiote to the surrounding environment.

If you've ever driven the stunning Pacific Coast Highway between Los Angeles and San Francisco, you might have an inkling of what Seely Brown is saying: Jaw-dropping cliffs give way to ocean vistas. Magnificently tall pine trees ring sections of the road. Surf crashes loudly ashore. It's one of the most serene drives in the world. Rolling down the windows brings in the best scents of nature: salt-tinged air mixed with sweet pine made more fragrant by the warm glow of California sunshine. It's easy to feel dreamy.

But it's also easy to feel disconnected from it if you're driving a boring old car that doesn't meld into the road the way a high-tech Porsche does. When your car isn't as attuned to its environment, you spend more time concentrating on the winding road and the mundane details of driving, rather than absorbing the experience itself. The more of the act of driving that you can off-load to the machine around you—in effect, the closer you can get to becoming a passenger—the more you can enjoy the serenity.

The Porsche exemplifies "ubiquitous computing," a term Seely Brown and fellow computer scientist Mark Weiser coined in 1988 while working at the Xerox Palo Alto Research Center (PARC). The duo mapped out computing's future and how it would flow from our desktops into the environment around us. They envisioned a world where computers make their way into everything, from the walls in our homes and street-side poles to our cars and clothes. The computers would be networked so one machine could call on the resources and power of every other machine in the system.

Science fiction writer Arthur C. Clarke famously said that the most advanced technologies are indistinguishable from magic; in much the same way, this new age of computing would be invisible. An addendum in the Xerox PARC predictions was that technology would also understand innately how humans work. In the two prior eras of computing we've experienced so far—mainframe and PC—the opposite held true. In the past, people had to adjust to machines and their awkward forms of input, whether with punch cards or keyboards and mice. That's unthinkable in the age of ubiquitous computing, where everything flows and answers to us. Nearly three decades after the concept was first articulated, at least one of its originators has deemed that the automobile and its elegant simplicity conform to that ideal. The humble car—well, the humble luxury car, anyway—has become a platform for computers that invisibly analyze their surroundings and adjust to them, enhancing and enriching the experience for its driver. This line between man and machine will only blur more as cars learn to drive themselves, a reality that major automakers such as Nissan believe will hit the mainstream by 2020.

But what about everything else? Before he died in 1999, Weiser said that we were entering this epoch. Most thinkers on the subject, including Seely Brown, believe we're firmly in the middle of it now because all of the pieces are in place, starting with ubiquitous computers. The number of personal computers and smartphones in use is expected to surpass four billion in 2015, double the number from the beginning of the decade.[3] In only a few more years, computers will outnumber the human population. If we count the myriad other devices powered by them—from Wi-Fi–connected digital bathroom scales and home thermostats to iPods and sports cars—then humanity was outnumbered a long time ago.

The networks that enable all these computers to connect and talk to one another are improving and expanding quickly, and not just in the developed world. People in the poorest countries are leapfrogging over PC-based computing and moving right to mobile devices that connect via cellular wireless networks. Even in countries such as Sudan and Papua New Guinea, the majority of the population is already connected. Pretty soon, every man, woman, and child will have a powerful, networked computer in the palm of his or her hand and all around as well. As we'll see in later chapters, the advancement of this connective technology will eradicate third-world poverty, change the global economy, and penetrate into the deep extremities of the planet.

The Internet of Things, where these connected machines supply each other with data directly without the need for user intervention, is also advancing quickly. Those bathroom scales and home thermostats already talk to our smartphones and in some cases think for themselves. (The Nest thermostat, for example, turns itself on and off based on its users' patterns.) The more that all of these ubiquitous computers communicate with one another, the more they can do so without human input. Where human control is necessary, more natural forms of interaction—such as touch screens, voice commands, and gesture control—have emerged to displace the often unwieldy keyboard-and-mouse duo. Not only are computers everywhere, they're getting easier to interact with.

Their ability to understand us is also getting better. One of the main ongoing projects at Google is teaching machines context or how to transform what they learn about us into useful information that can be applied at the proper time and place. "Without an understanding of what we want to know and when we want to know it, it's difficult for systems to achieve their potential. We're just beginning to be able to do that effectively," says Alfred Spector, Google's vice president of research and special initiatives. "No one gets it right. Even our best friends misread our moods sometimes. That's among the most important challenges."[4]

If you teach a computer ten facts about yourself, the machine can provide you with regular updates that you likely will find interesting. If you've ever shopped online and a website suggested new items for you based on past purchases, you've seen this in action. But how does the machine know *when* to tell you? After all, not many people want to have an important conversation interrupted to learn that their favorite band has just announced a tour. Nor do you want Amazon waking you up in the middle of the night to let you know that the latest season of the TV show you're obsessed with is available on Blu-ray. (Well, most people don't.)

In 2012, the search engine company introduced Google Now, an "intelligent personal assistant" for smartphones. The system predicts what a user wants based on how it has been used in the past, then actively delivers the information to the user. If you have a meeting scheduled, for example, your phone will calculate directions, check traffic, then alert you about when you should leave based on that information. Pretty soon, our phones and other gadgets—like our luxury cars now—will monitor and analyze their environment and user, intelligently and autonomously anticipate our needs, then suggest courses of action.

Bill Buxton, another veteran of ubiquitous computing from Xerox PARC and now a principal researcher at Microsoft Research, gets as philosophical as Seely Brown when describing the current era. He believes we are standing on the cusp of societal transformation, with the melting of technology into our surroundings and our subsequent mastery of the environment now just a matter of execution and details. "This is the best time ever because our dreams are within our reach now. It's just technique now. The physics aren't in the way. We've got the paint, we just need to figure out the perspective so we can create the masterpiece, so to speak."[5]

About thirty miles away from the Pacific Coast Highway lies a school campus that wouldn't look out of place anywhere in North America, except for the joint civilian-military airfield behind it. The mottos of Singularity University—"Preparing humanity for accelerating technological change" and "Solving humanity's grand challenges"—sound as bold and ambitious as its curriculum. Its core courses focus on the most important emerging technologies, from robotics and artificial intelligence to neuroscience and nanotechnology. The faculty and students also distinguish Singularity U. from most other schools. The list of instructors and lecturers reads like a who's who of the technology world, including human genome pioneer Craig Venter and original Internet architect Vinton Cerf. The student body consists of chief executive officers, venture capitalists, and politicians.

With funding from NASA, Google, and several other technology companies, futurist Ray Kurzweil and engineer-entrepreneur Peter Diamandis founded Singularity University in 2008. It's not so much a school as a think tank or an exclusive club where big thinkers and power brokers meet and swap ideas in preparation for a future in which rapidly advancing, increasingly aware technology surrounds us. "We're pulling in the future CEOs and university presidents and government ministers when they're young in their careers," says Diamandis, "pulling them together and allowing them to really meet in a setting where the message is, 'Anything is possible, what is the future?'"[6]

Of particular concern to Singulartarians is the approaching turning point in human evolution where ubiquitous computers surpass human intelligence—that inevitable direction in which Moore's Law is pointing us. When measured by millions of instructions per second (MIPS), computers surpassed the raw calculating power of insect brains somewhere around the year 2000 and mice brains around 2010. Estimates vary as to the computing power of the human brain, but several agree on around one hundred million MIPS. Watson, the IBM supercomputer that became a *Jeopardy* game show champion in 2011, could perform about eighty trillion operations per second or about eighty million MIPS. Using historical growth rates as an indicator, experts such as Carnegie Mellon robotics professor Hans Moravec believe the average desktop computer will surpass the human brain's raw computational capability sometime in the 2020s.[7] Of course, MIPS deal only with basic, mechanical functions such as walking, breathing, and seeing. Tracking a white spot against a mottled background, for example, requires about one MIPS, while finding three-dimensional objects in clutter calls for at least ten thousand

MIPS.[8] Those are interesting figures but a far cry from true intelligence and an even further cry from wisdom in the human sense.

IBM's Watson and Google Now represent early steps toward that other side of the equation, the wisdom part. While hardware capability is doubling every year or eighteen months, algorithms are getting better and cheaper even faster. Just a few years ago, the Microsoft Office program suite cost several hundred dollars; now equally powerful and continually improving alternatives to it are available for free. This software improvement, more than hardware acceleration, is fueling machine intelligence—and the knowledge of what to do with it—by enabling smart personal assistants, trivia mastery, and other metrics. It's also what's driving technology into the background environment and making it aware of our human context. None of this is happening overnight, though, which is why Kurzweil—sometimes referred to as the High Priest of the Singularity—thinks it will take machines a while longer to surpass actual human intelligence. But it won't be that much longer; he thinks it will happen by 2045.[9]

The timing is up for debate, but the event itself looks inevitable. Matters will get really wild after that because biological human intelligence will continue to grow with the same linearity as it has throughout our history, but machine intelligence will continue to multiply exponentially. Soon after this singularity event, a single computer will outstrip all of human intelligence combined. After that point, a simple traffic light or cellphone—if we're still using either of them as we know them now—will be smarter than the entire human race.

It sounds like scary science fiction, but it isn't. The concept of singularity comes from astrophysics, where it refers to a point in space-time where the rules of ordinary physics cease to apply. In a real-world sense, it refers to a future hard to envision because nothing like it has come before. But humanity has experienced several singularities already, the printing press representing perhaps the best example. Prior to Gutenberg's invention in the 1450s, the act of reading, and therefore the absorption of information, was accessible only to a few privileged members of society. The majority of those elites, never mind the general public, couldn't realistically envision what would happen when knowledge became readily available to the masses. Fears and speculations abounded, just as they do now, but no one really knew for sure. In retrospect, the world changed dramatically and for the better as the population's collective intelligence expanded and new ideas spread rapidly. Still, the average person living in the fifteenth century couldn't possibly have imagined robots on Mars or that he or she someday would have access to the entire catalog of human cultural creation through a hand-held device.

We're in the middle of a similar situation now, with the Internet heralding a similar sea change. In 1969, when Cerf and his team at UCLA connected the first link of the ARPAnet, the military precursor to the Internet, they had no idea what forces they were unleashing. Their creation went on to transform information transmission and communications, topple governments, destroy industries, and create new ones. It also greatly simplified the sharing of cute cat videos.

In the late nineties, just as the Internet was taking root, journalists were still assembling daily newspapers by printing out stories and headlines on paper. We cut them up and glued them onto larger pieces of paper, which then went to a large-scale printer. Reporters telephoned people to confirm important details in their stories, such as where a company was based or the spelling of a person's name, then ran over to the newspaper's physical library to check facts in books and older issues. Nowadays, the Internet has simplified and replaced just about the whole process. It's a bit like a reverse Singularity effect—just as I couldn't have imagined then what a journalist's job would be like now, I can't really recall how things were possible then. Similarly, it's difficult for just about anyone in a developed country now to consider that people had to draw water from holes in the ground.

One of the first references to the word "Singularity" in a technological sense came in 1965, when British mathematician I. J. Good described what would follow the invention of a super-intelligent computer or one smarter than humans: "An ultra-intelligent machine could design even better machines; there would then unquestionably be an 'intelligence explosion,' and the intelligence of man would be left far behind. Thus the first ultra-intelligent machine is the last invention that man need ever make."[10] This ultra-intelligent machine has fueled fears, some stoked by popular science fiction, about a future in which humans are irrelevant, but as we'll see that's not going to happen. Every step toward automation in the past has led humans to new and previously unimagined heights, which gives us no reason to believe this time will be any different despite the understandable fear of the unknown.

Ubiquitous computing and the approaching computational Singularity mean we're entering a new age of human evolution. The first humans had only primitive technology and were largely subservient to their biology and environment. Their descendants—the second humans, us—fared considerably better by using technology to coexist with nature. Now, we're in a third era that geologists are calling the Anthropocene epoch, in which humanity, through technology, has become the main determinant of all of the world's systems, including biology and environment. While the era's official start date is under debate, the consensus holds that it applies to the point when humanity started to influence the whole planet in a meaningful way. Some peg it to the start of the Industrial Revolution in the eighteenth century, when technology finally gave us the ability to manipulate, enhance, and damage nature on a large scale; others think it started more recently, with the arrival of digital technology and the Internet. Whenever it began, we're in the thick of it now.

It turns out that the X-Men, with all their fabulous superpowers, are coming after all but through technology rather than biology. We may not have biological superpowers, but with external technology we can fly across oceans in hours, instantly look up any fact, cook dinner remotely with our phones, drive a car without touching the steering wheel, all of which would seem magical to any of our ancestors. Anthropologists might call this next step *Homo technologicus* or *Homo superior*, but Latin seems outlandishly anachronistic given the context. It seems more fitting now to adopt the language of technology when describing the new epoch since it's such a key part of it.

We're not evolving, we're upgrading, just like software. In the third age of humans, as people master nature rather than coexisting with it or being subjugated by it, we are becoming Humans 3.0.

## HAIRDO 2.0

In the world of technology, moving from one software version to the next generally means an improvement of some sort: fixing bugs, new features, or an updated interface. That isn't always the case though. Plenty of examples of the opposite exist. A case in point: the great Windows Vista debacle of 2007. In an effort to make its operating system more secure, Microsoft introduced a host of features with Vista that bogged computers down and made them slower and more annoying to use. Many angry users reverted to Windows XP, an older yet simpler version of the software. The company largely fixed the problems with the well-received Windows 7 in 2009, but they alienated users again in 2012 with Windows 8 and its interface completely redesigned for mobile devices.

Where do Humans 3.0 fall on this spectrum? Are we becoming better through this upgrade we're experiencing, or are we the biological equivalent of Windows Vista or Windows 8, new but not necessarily improved? Analogies aside, this is perhaps the most prudent question of our time: If technology isn't making the world a better place and improving humanity's lot, what is the point of it all?

Ask anyone whether people are better or worse off because of technology, and you'll get a complete spectrum of answers. We'll hear lots of these opinions in the chapters ahead. Some take a dim view, that, while technology has solved a lot of problems, it has created even bigger ones. Others say that technology has moved us further away from what makes us human, that we are becoming more like machines. Still others disagree, saying that we're nearing the sort of technologically driven utopia promised back in the idealistic 1950s.

Before I started this book, my views fell somewhere in the middle. I'm no techno-utopian, nor did I start my career as a fan of technology. In fact, as a long-haired teen in high school, I thought I'd grow up to be a journalist who interviewed rock stars for a living. But then the 1990s happened, and reality set in. All my grunge idols cut their hair, and the prospect of needing steady employment crept closer and closer. As every young person learns, companies don't hand out dream jobs at graduation, especially to long-haired freaky people.

So I ended up a short-haired, low-paid writer at *Computer Dealer News*, a trade magazine. I knew virtually nothing about technology, but I was part of the first generation of kids who grew up in this emergent ubiquitous computing epoch. Understanding the lingo and the beat came relatively easily once I dived into it. The only story I can recall writing at that job was a feature on why uninterruptible power supplies—or the boxes that prevent computers from frying during an electricity outage—were important. We don't need to get into the details, but trust me: They're important.

Many people who write about technology do so because they have some kind of background in the field; they're programmers on the side, or they have engineering degrees and somehow fell into writing. As such, a lot of technology news focuses on—even obsesses over—"speeds and feeds." If the stories aren't about the latest specifications of the newest gizmos, they're about the gadgets themselves. In recent years, an entire sub-industry of technology journalism has sprung up, dedicated to *rumors* about new gizmos and their specs. Obviously an audience exists for that kind of content, celebrity gossip for nerds, but much of it is speculation and hype devoid of context.

In my previous book, *Sex, Bombs, and Burgers*, I developed a deeper appreciation for what technology really does and what it means to society. That book took the premise that three of the worst-regarded industries on the planet were in fact responsible for many of the most amazing and important technological discoveries and advancements. Military research has contributed to an untold number of deaths over the course of history, but it has also led to incredibly important and positive developments, such as x-rays, the Internet, and space exploration. Pornography purveyors, meanwhile, have a bad rap for exploiting their workers and loosening society's morals, yet they also quickly adopt and invest in new communications technologies—including the Internet—when no one else is willing to take the risk. Fast-food processors and restaurants have caught grief for contributing to poor nutrition and obesity, but they receive almost no recognition for their vital roles in securing and stabilizing food systems, the very bedrock of modern society.

Writing *Sex, Bombs, and Burgers* gave me a new perspective on technology. It's not innately positive or negative: It starts out neutral, but even its negatives have a tendency to turn positive. Any new development in most any field usually has undesirable side effects, but those become novel problems that result from the solution of old ones. The new issues usually aren't as big or serious as the old ones; if they were, the public wouldn't accept the technology behind them. Cellphones stand as a good example. Some say we've become addicted to them and that continually staring into our palms means we're no longer conscious of the world or the people around us. That's certainly true in some cases, but the overall benefits of cellphones to the larger population are obvious: They've given us a newfound freedom to obtain information and communicate with one another. In the developing world, they're even more important. They allow millions of people to communicate, educate themselves, and engage in commerce for the first time ever. In nations with state-controlled media, they also allow for the transmission of politically sensitive or dangerous information among the people. Overall, that's a good tradeoff, which is why people have taken to them by the billions.

Exceptions apply to the rule of course. Technological progress has brought about several existential crises, such as the possibility of nuclear annihilation. The threat has died down since the end of the Cold War, but it still lingers. Large-scale devastation has also become possible by other technological means, such as the escape of a bioengineered virus or rogue nanotechnology. There is also the possibility that global warming, if it continues unabated, will wreak havoc on all aspects of life. These big problems need solutions that don't incur even bigger ones, but humanity's track record is good so far, given that we're still here. These threats are not to be taken lightly, but we do have reason to be optimistic.

Consider the population explosion that took place as the result of the Industrial Revolution: With better sanitation and health practices, infant mortality rates plummeted, and people began living longer, which gave rise to the threat of the world running out of food and other resources. Famine and scarcity of key supplies still exist, but not on the levels feared because other technological advancements, especially in food production, mitigated the negatives.

Ironically, writing a book about humanity's negative traits ultimately gave me a more optimistic outlook on our trajectory and about technology in general. In that vein, I believe there's an empirical way to answer the question of whether technology is good for us and to divine whether it will continue to make us better in the future. First we need to look at what it means to be human, the various aspects of who we are. The jobs we perform, how we spend our leisure time, the way we communicate and relate to each other, our beliefs—these are all part of the human experience, and fortunately they can be quantified, measured, and compared with the past to form a baseline that allows us to peer into the future.

Pragmatists say it's folly to try to predict the future, but technologists have been forecasting events accurately for decades using Moore's Law. The digital camera is a great example. Steve Sasson assembled the first one in 1975 while working at Kodak. It successfully captured images digitally rather than by using film, but it was big, ugly, and expensive. Applying Moore's Law to its various components—the processor, image sensor, memory storage, and so on—he guessed that it would take fifteen to twenty years for his creation to evolve into something good and cheap enough to be commercially viable. The first digital camera hit the market in 1994, nineteen years after its initial invention. "The prediction actually turned out to be not too far wrong," Sasson says, putting it mildly.[11]

Apple produced a promotional video in 1987 in which a professor prepares for his lessons on a flat, tablet-like computer. Most of his interaction with the futuristic device takes place through a "knowledge navigator," a voice-controlled assistant that looks through his files and performs certain tasks, such as launching a video call. The date in the video is 2011, the same year—amazingly—that Apple launched its Siri voice assistant and a year after it released the first iPad. The video may have seemed like fanciful science fiction back then, but it represents another good example of how Moore's Law can predict how the future will look.

Predictions like that are forming every day in labs, tinkers' basements, and boardrooms around the world. When a new technology appears, it usually doesn't work very well, it isn't conveniently sized or visually appealing, and it's extraordinarily expensive. All of these issues lessen, however, as the technology's ecosystem evolves and improves. The hard part is the breakthrough itself; the rest is usually only a matter of time. That's what allows technologists and engineers to design software, for example. They can guess with a relatively good degree of accuracy what hardware specifications will be in a few years' time, so they adjust plans and expectations accordingly.

Outside of technology, historians, sociologists, and economists have been predicting the future for some time by examining past trends. Much as with Moore's Law, they apply their own analytical formulas to make educated guesses on the trajectories of metrics such as economic growth, demographics, and even political changes. In the late nineteenth and early twentieth centuries, German philosophers Georg Wilhelm Friedrich Hegel and Oswald Spengler examined past events to predict and explain the rise and fall of societies, while others such as University of Connecticut professor Peter Turchin more recently have charted prior episodes of violence to predict upcoming ones. According to Turchin's models, for example, America is on course for major civil strife because of growing inequality, a cycle we have seen before.

Such waves of social pressure are inevitable, though they can be ameliorated with preventative steps, which is the point of charting them. The United States averted revolution in the early twentieth century through progressive government policies that included regulatory crackdowns on corporations and new rights for workers.[12] A nuanced understanding of the past can help us foresee major future events or trends.

From the perspective of Moore's Law, it's hard not to conclude that our destiny is taking us down a certain path. Going from the discovery of fire and the creation of the wheel to walking on the moon and charting the genetic code of all living creatures in just a few thousand years isn't just astonishing, it reveals the nature and future of our species. Inevitable hiccups and detours will happen along the way, but we are marching inexorably toward more understanding and intelligence. Whether we're acquiring wisdom along the way remains an open question, but the future looks inevitable. On a long enough timeline, humanity has the potential to arrive at total intelligence and total understanding of the universe. The real question isn't whether this is happening but rather how long it will take and what happens when we get there.

There is value in examining the direction of technology because it affects, influences, and makes possible almost all aspects of human civilization. As we'll see, technology drives prosperity, health, and longer life. It dictates how people spend their time in the jobs they work and the leisure activities they enjoy. It determines how people communicate and how much space they have—both physical and virtual—between one another. Technology also shapes peoples' identities and beliefs. Taken together, these aspects of life determine who people are both individually and collectively and ultimately whether they are content with their place in the universe.

So where is technology taking us? Arriving at total understanding will change how we look back at the past—that is, where we are now—so let's look forward from here and find out what's happening to us along the way. The chapters ahead attempt to quantify, measure, and compare what it means to be human over the time that we've been on Earth. Hopefully you'll agree that Humans 3.0 are a new and improved species.

# 2

# Economics: Widgets Are Like the Avengers

*In the multitude of the people is the king's honor; but in the want of people is the destruction of the prince.*
—PROVERBS 14:28

Technology's dramatic effect on humanity becomes most obvious when it comes to the economy, which makes that a good place to start our journey. The economy forms the foundation of our society, with everything else built on top of it. Innovation drives the economy since it enables us to divert our attention from menial tasks to newer and bigger endeavors. That's why technology companies jingoistically refer to their products and services as "solutions." The best of them do solve problems.

Let's go back to the very beginning, to Blarg, a fictional caveman. Before innovation or "solutions," Blarg's life was tough. He shivered naked in a cave, subsisting on berries and roots that he managed to scavenge. When he wasn't calling in sick to work because of pneumonia, he was doubling over at home with stomach pains. It wasn't the most productive of times. But then our intrepid caveman discovered fire. Suddenly, he had a way to heat his hovel and even cook the flesh of small animals. Plus, he could see who was trying to grope him in the dark. No more *E. coli* or unexpected nighttime dalliances.

With the consequent improvements to his physical and mental health, Blarg took fewer sick days, which meant he and his tribe increased their daily berry and root harvest. This boost to surplus meant they could now trade with a nearby tribe for furs, which meant they didn't have to shiver naked anymore. Or they just scared the hell out of their neighbors with their newfangled weapon and took said furs. Either way, fire improved human productivity. It also probably sparked an arms race, with neighboring tribes working on producing bigger and better fire, which became one of the first "solutions."

The wheel, invented thousands of years later by Blarg's descendants, also gave a huge boost to economic development. Not only could it help spin clay and create pots, it turned rapidly under a waterfall. In both cases, villagers made pottery and churned grindstones faster. When the wheel made its way into vehicles such as horse-drawn carts, travel and bearing heavy loads across long distances became possible. In all these applications, simple innovation eased the amount of work humans had to do in order to achieve the same or even increased results. Our backs have been better for it since.

Fast forward to the Age of Exploration, when international trade expanded in search of new trade routes and products abroad. Key to that endeavor was the invention of two devices we now consider extraordinarily mundane: the compass and the hourglass. GPS has rendered the former essentially obsolete unless you're a wilderness scout, while the latter matters only to soap opera fans who watch *Days of Our Lives*. Yet by giving navigators the ability to time their journeys and direct them away from hazards such as bad weather, rocks, and mythical creatures like the Kraken, compasses and hourglasses doubled their effectiveness. Ships using such gadgets could make two round-trip journeys a year from Venice to Alexandria, for example, rather than just one. Other technological improvements, such as better materials and stronger structures, also meant that with a much smaller crew an eighteenth-century ship could carry ten times the cargo of its fourteenth-century predecessor.[1]

Throughout this brief timeline of human history, technological advances also spurred competition between individuals and entities. Fire may have given the advantage to one caveman tribe looking to steal furs, and better navigational tools the advantage to one colonial empire over another. Technology literally became the secret weapon. In more modern times, technological supremacy also determined global hierarchies. The twentieth century belonged to the United States because it first mastered and institutionalized scientific research and then combined it with close ties among government, businesses, and education—as seen in the development of the atomic bomb during the final years of World War II.[2]

Silicon Valley, the current hotbed of international technology, has played a dual role since its inception: military incubator and steady source of consumer innovations. We know Google as a search engine company, but we don't hear nearly as much about the work it does for the military in fields such as robotics and analytics. Much of Google Earth and Google Maps, the omnipresent phone apps, were developed by Keyhole, a CIA-backed company purchased by Google in 2004. The company's ongoing efforts in robotics are taking place in conjunction with the Pentagon. In 2013, it purchased Boston Dynamics, the contractor responsible for military robots such as the four-legged pack horse, BigDog, and the humanoid rescue machine, Atlas.

With the benefits of technology—increasing efficiency and productivity, fueling competition, and opening new markets—the world's economy has grown exponentially. Economists estimate the worldwide gross domestic product (GDP) in the year zero at about 102 billion international dollars, the fictional currency used to measure values among places and times. Toward the end of the nineteenth century, during the golden age of shipping, that figure had climbed nearly a thousand percent to Int$1.1 trillion. By the close of the twentieth century, global GDP rose a further 3,000 percent to Int$33 trillion.[3] That's 33,000 percent growth since the reign of the Roman emperor Augustus. Even more surprising is that, despite all the doom and gloom in recent years about economic downturns, fiscal cliffs, and bailouts, there's no sign of the world's overall economy slowing down any time soon. The world's GDP in 2013 was around $70 trillion, almost double what it had been only fifteen years prior.[4]

Increased globalization explains this acceleration. After the Cold War ended without a disastrous nuclear exchange, most of us found it immensely beneficial to cooperate with one another. Previously suppressed markets behind the Iron Curtain suddenly opened to North American and Western European business. More importantly, the détente made trade easier with China and India, two huge markets over which Western countries had been salivating for decades. On both sides, more people were free to participate in the new global economy than ever before.

Just as in the age of sailing, technology drove the growth. The Internet, which arose at the same time as this newfound harmony, provided the perfect tool at the perfect time. In the Western world, it became the ultimate equalizer that allowed entrepreneurs to compete with big, established businesses on an even footing, eventually leading to new wealth emerging from nothing. It also tied the East to that growth, bringing a number of Asian countries along for the ride.

On the one hand, the Internet enabled the rise of companies such as Amazon, eBay, Facebook, and Google, a quartet that took fewer than twenty years to amass a combined market capitalization greater than the big six automakers had in a century.[5] On the other hand, it also allowed traditional businesses from AT&T to Shell to avail themselves of the East's cheaper labor through Internet-connected call centers. Manufacturers and retailers, meanwhile, cut costs by moving production to China. Using just-in-time delivery technology, companies such as Dell and Wal-Mart introduced hugely profitable efficiencies into their businesses.

The West got richer, but the East may have benefited even more. China and India posted stellar economic growth numbers during this period. Much of that filtered down to the general populace, resulting in huge reductions in overall poverty. In one of the great unreported stories of our time, the proportion of people living in extreme poverty in the world fell by half between 1990 and 2010, beating the United Nations' Millennium Development Goal by five years. Much of this happened because of the immense growth in China and India but also despite the global financial crisis. Nor was it just an Asian story: Outside China, the poverty rate fell from 41 percent in 1981 to 25 percent in 2008.[6]

The Internet has brought about huge economic benefits wherever people have steady access to it. A 2011 study of the G7 nations plus Russia, Brazil, China, India, South Korea, and Sweden found that the Internet contributed about 3.4 percent annually to GDP in those countries, an amount equal to the entire economic output of Spain or Canada. Internet-related consumption and expenditure now exceed agriculture or energy.[7] As a tool, it has fired the imaginations of entrepreneurs the world over and introduced tremendous efficiencies to existing businesses. Moreover, the bigger impact has come from its ability to act as an instrument of creative destruction, disrupting traditional ways of creating, producing, and distributing products and services and replacing them with better methods, which we'll see in later chapters.

With the Internet revolution only just beginning, the future is set for further exponential growth. Fledgling fields such as robotics, nanotechnology, genetics, and neuroscience represent entirely new markets for businesses and individuals to cultivate, while emerging markets in South America and Africa will ensure that the pie keeps getting bigger. Forecasters expect the global economy to double again to around $143 trillion by 2030.[8] Leading the way will be the so-called emergent seven (E7) countries: China, India, Brazil, Mexico, Russia, Indonesia, and Turkey. Their combined economies now represent about 70 percent of the G7, but by 2050 they will be double its size.[9] The balance of power is shifting and the world order as defined by population, where the size of a country's market becomes its key influencer in global affairs, is balancing out. Back in the year zero, China and India accounted for a combined 70 percent of the world's economy.[10]

This isn't the end of the figurative world for the West because it's not a zero-sum game. It's actually a win-win situation, with smart, forward-thinking Western businesses benefiting from the opening and maturing of additional markets. As a PriceWater-houseCoopers report puts it: "This larger global market should allow businesses in G7 economies to specialize more closely in their areas of comparative advantage, both at home and overseas, while G7 consumers continue to benefit from low-cost imports from the E7 and other emerging economies."[11] While already developed nations will see their overall share of global GDP shrink, their populations will still enjoy comparatively high income for some time. Chinese workers, for example, likely will earn only half the salaries of their American counterparts by 2050, while those in India will still make around only a quarter.[12] These countries are advancing quickly and gaining more clout in global decision-making, but they still have hundreds of years of economic development to match.

When viewed through the detached lens of history, the long-term economic future of the world shines brightly, despite what we hear daily in the news. It's no wonder that some experts are upbeat, including Nobel Prize–winning economist Edward Prescott, who has proclaimed that "the whole world's going to be rich by the end of this century."[13] Microsoft founder Bill Gates is even more bullish: "I am optimistic enough about this that I am willing to make a prediction. By 2035, there will be almost no poor countries left in the world."[14]

## GINI IN A BOTTLE

But what about Occupy Wall Street and the decimation of the middle class? If everything's going so well, why did thousands of people camp out in city parks around the globe a few years ago? Weren't they saying the opposite, that the world is going to hell?

The movement had many messages, but, if the protesters' central argument opposed the growing gap between rich and poor, they weren't wrong. Global economic growth has accelerated dramatically over the past few decades, but so too have income disparities. The countries of the world have made significant progress toward equality among one another, but internally many have been heading in the opposite direction for several decades. Just as with technology, the world's runaway economic growth also has a downside.

The Kuznets Curve—coined by Harvard economist Simon Kuznets in the mid-twentieth century—holds that inequality, in the shape of an upside down U, ties to industrialization. During a country's early modernizing stages, inequality rises as farmers leave their land for higher-earning factory jobs in the city, creating an income gulf with those who stay behind. But as more people get better educations, they demand more equitable income redistribution from their government, which ultimately leads to a decline in inequality. That demand can take the form of organized opposition and even revolt. The same force gave birth to communism in the nineteenth century and unionization in America in the early twentieth.

The Kuznets Curve held its shape until around 1980, when it became more of a capital N, whereby inequality in many nations began a sharp turn upward.[15] A variable called the Gini coefficient, coined by Italian statistician Corrado Gini in 1912, measures economic inequality. A rating of zero represents total equality: All of its people share a country's wealth evenly. A score of one means the opposite: complete inequality, where one individual gets everything. No country—not even the most socialist welfare states or the most despotic tyrannies—is likely to hit either end of the spectrum, but the range of the scale means that even a small uptick represents a big increase in inequality. America has seen a lot of these small upticks, gaining nearly a tenth of a point over the past three decades, moving from 0.3 around 1980 to almost 0.4 in

2010. In more concrete terms, that means the disparity in disposable income between the top and bottom earners has climbed by almost 30 percent. China's gulf has grown even bigger, around 50 percent. Conversely, the Gini coefficient in South America, traditionally the world's most unequal continent, has fallen sharply over the past decade for reasons we'll see later. Otherwise, the majority of people live in countries in which income disparities have grown bigger than they were a generation ago.[16]

The reason is a paradox: The same globalization that has enabled poorer countries to improve their lots by joining the world economy also has exacerbated income disparities between individuals by allowing the rich to get much richer. Thirty years ago, if you started a company in the United States and wanted to sell widgets, most of your customers would be in America. Maybe you could sell those widgets to Canada since it's close by. You could have become rich doing this, and many people did. Still, that total market had somewhere around 350 million potential customers.

In a newly opened world, that same potential market has expanded into billions of potential customers. Smart business owners capitalized and became richer than they could have if they had stuck only to their home markets. Movies offer a great example. *E.T.: The Extra-Terrestrial* was the highest-grossing film of the 1980s, pulling in $435 million at the box office. Marvel's *The Avengers* raked in a comparatively astronomical $1.5 billion in 2012, adjusted for inflation. Nearly 60 percent of that haul came from non-US theaters.[17] That figure results both from globalization and from people in developing economies having more disposable income.

On the plus side for the individual, globalization has changed the nature of wealth creation. In the past, the most likely way to riches was through inheritance or owning land, but now plucky individuals can get there through entrepreneurialism or working at a hot company with stock options. Merit and opportunity have reentered the picture.

But inequality remains a big downside for those not partaking of those new opportunities. Exacerbating the issue—and a particular bugbear for the Occupy protesters—is cronyism, the cozy relationships that exist between politicians and the business world. It's particularly bad in Russia, China, and India, which have a disproportionate number of billionaires thanks to insider access to land, natural resources, and government contracts. But it's a problem in Western countries too, where government efforts to curtail harmful corporate behavior—regulating banks, oil companies, or telecom firms—are often lobbied into submission.

In real terms, all of this means that lower- and middle-class incomes in many developed countries have grown more slowly or not at all while the upper echelon has skyrocketed. In 2012 alone, the number of billionaires in the world increased by 10 percent to 2,160, while their combined wealth grew 14 percent to $6.2 trillion.[18] No wonder people got angry.

A growing body of evidence shows that income disparity harms society. It leads to greater crime, worse public health, and other social ills, not to mention greater stress on people, which can result in high blood pressure, weaker immune systems, and even decreased sex drives. Income-related stress even affects the brains of children; one study found that seventeen-year-olds in poverty could circulate about eight thoughts at once, compared to around nine for better-off teens.[19] One extra thought seems minor, but it could contain the seed that becomes the next Google. Never mind the mind: A thought is a terrible thing to waste.

Thomas McDade, a biological anthropologist at Northwestern University, says the effects of income disparity run deep:

*We're coming to understand that even if you have a stable job and a middle-class income, then your health is not as good as that of someone who is in the one percent. There is something more fundamental about social stratification that matters to health and the quality of social relationships. It matters because of what it means: can I participate in society?*[20]

We'll return to some of the social effects of inequality later, but for now there is some good news: We've already seen this movie several times. As the folks on *Battlestar Galactica* are fond of saying: All of this has happened before and will happen again.

Governments generally can address and fix growing inequality in three ways: taxes, spending, and regulation. Using all three properly is usually the only way to avoid a popular uprising. (See the French Revolution for how not to do it.) Rising inequality in Germany led Otto von Bismarck to introduce pensions and unemployment insurance in the 1880s while Theodore Roosevelt's Square Deal in the early twentieth century broke the monopolies abusing workers and fattening business owners. Unions became powerful in the aftermath of those reforms, and minimum wages protected by law helped narrow the rising gap. While many European countries have opted for stronger redistribution of wealth through higher taxes (for example: tax-happy Sweden has the world's lowest Gini coefficient, 0.24), North America historically has emphasized equality of opportunity, the ability of a person born to few or no advantages to climb the ladder into the echelon where he or she can take baths in Cristal and buy diamond-studded toothbrushes. The best way to ensure opportunity, economists agree, is for governments to invest in education—and not just universities, since that just reinforces class division, but at all levels.

US inequality shrank considerably between World War II and the 1970s, precisely because of that investment in education. The G.I. Bill, a package of benefits available to troops after the war, resulted in millions being able to afford higher education and home mortgages, thereby boosting their chances at upward mobility. But since then, that mobility has plummeted. In the 1970s, one year's tuition at a public university cost about 4 percent of a typical household's annual income, a number that jumped to 10 percent by 2009, thereby pricing many students out of postsecondary education. America is almost unique among wealthy countries in that men between the ages of twenty-five and thirty-four are now *less* likely than their fathers to have a college degree.[21]

The question then becomes: How long are the governments of the world going to allow this inequality to keep growing?

## JOKER'S DILEMMA

Global economic growth and the internal inequality that comes with it bring to mind the Prisoner's Dilemma, a manifestation of game theory that became relevant during the Cold War. We can consider this brewing discontent between the rich and the not-rich a new type of conflict—let's call it the Boiling War.

With tension between superpowers at a new high following the Soviet Union's detonation of an atomic bomb in 1949, the US government charged the RAND Corporation with thinking the unthinkable: how to avoid the mutually assured destruction inevitable in a nuclear exchange. In 1950, Merrill Flood and Melvin Dresher, a pair of mathematicians working at the American military think tank, came up with a theoretical game that perfectly illustrated the dilemma. The test mapped the effects of two sides either betraying or cooperating with one another. Each result, however, meant unappealing consequences for all participants.

A third mathematician, Albert Tucker, refined the concept later that year by giving it a more tangible if less apocalyptic form. In his permutation—still used in just about every cop show today—police have arrested two men but don't have enough evidence to convict either. The law enforcement officials separate the two suspects and make the same offer: If one betrays the other and testifies against him, the snitch goes free while the other guy goes to the clink. If neither talks, both men get a month in jail. If both rat on each other, they each get several months. The police leave the men to stew. The Prisoner's Dilemma, as it has come to be known, has been tested, debated, and fictionalized ad nauseam in various forms ever since.[22] Even Batman got in on the action in *The Dark Knight* when the Joker offered separate boatloads of civilians and criminals the choice of blowing up each other. If neither acted, both would be vaporized. (Clearly the Joker had boned up on Cold War game theory.)

The essence of the dilemma is that participants must choose the outcome most beneficial to themselves. In the case of Tucker's prisoners, the best choice seems to be to rat on the other guy and go free.

But unspoken and perhaps unknown repercussions apply to such a decision. What if the fellow suspect is a friend? What if he chooses to retaliate once he's free? In the case of the Joker's victims, it would have been easy for either boat to blow up the other, but their occupants would have had to live with those deaths on their respective consciences. That's not an obvious choice for anyone, including—as the movie suggests—hardened criminals. They're people too, gosh darn it.

Over time, betrayal has proven to be the optimal choice in theoretical tests of the Prisoner's Dilemma, but psychologists and social scientists have found non-immediate repercussion factors to be quite strong, to the point where real-life test subjects choose to go for cooperation.[23] When put to the test, people and the organizations they form are subconsciously far-sighted and tend to cooperate rather than betray one another.

The best proof we have of this fact is that the world is still here. We didn't blow ourselves up in a nuclear Armageddon years ago. The Soviet or American governments could have nuked their sworn enemies off the map, but what would the fallout be?

In its 2011 report on world trade, the United Nations boldly proclaimed that "economic integration and interdependence in the world today have reached an unprecedented level," a statement that comes as no surprise given the strong move toward globalization. Over a few centuries, our nature has shifted from zero-sum competition to win-win cooperation. Rather than competing exclusively for resources as in centuries past, the nations of the world are learning how to share those resources for the benefit of all. Countries are figuring out ways to trade their own surplus goods and services for what they need. People are no longer obstacles to an end, they are potential customers to whom products can be sold. Now, more than any other point in history, people are more valuable to each other alive than dead, which is the ultimate expression of cooperation conquering competition.

This is why people aren't killing one another as much as they used to. The evening news might lead you to think otherwise, but it's true. The size and scope of conflict-related death up until the middle of the twentieth century was astonishing. During this long chunk of human history, mass death was commonplace. Some fifty million people perished during the Muslim conquest of India in the first half of the second millennium AD. The Qing dynasty's conquest of the Ming in the seventeenth century led to the deaths of twenty-five million. The nineteenth century Taiping Rebellion had twenty million. World War I saw thirty-seven million deaths, and World War II tops all human conflicts, with an estimated death toll falling somewhere between fifty and seventy-five million.[24] Average those numbers out, and it's like losing the entire population of Argentina in one fell swoop. These numbers are unfathomable today.

Some agitating observers suggest that a war between America and China—one of the emerging superpowers—is inevitable, but it's improbable given how linked the economies of the two countries have become. Any conflict between the two that goes beyond the current cyber-espionage and possibly even sabotage would devastate both. The populace usually finds war abhorrent, and governments of developed or nearly developed countries generally come around to that way of thinking too, largely for economic reasons. The cooperative benevolence side of game theory is winning out.

The ongoing conflicts in which many advanced and developing countries find themselves today are often significantly different.

The protracted conflicts in Afghanistan and Iraq both started as state-versus-state battles, but with quick official "victories" they evolved into what military types like to call "asymmetrical warfare," more popularly known as insurgency or guerrilla warfare, depending on which side you're on. Asymmetrical warfare has its benefits as well as disadvantages. On the plus side, it results in a lot less death. Estimates place the total number of military and civilian deaths in the first ten years of both the Iraq and Afghanistan conflicts at around 140,000.[25] Any unnecessary death is one too many, but that's a far cry from the jaw-dropping tolls racked up in similar wars only a century earlier. No conflict outside Africa has racked up more than a million casualties since the end of the Cold War.

The unfortunate downside of asymmetrical warfare is that it's much more unpredictable, which means its definition can be expanded to include terrorism. In a traditional war, residents of participating countries sometimes have the option of fleeing. Guerrilla fighters, however, can and often do strike randomly, and terrorists generally go after defenseless targets such as civilians because attacking their real enemies would place them at a disadvantage. Innocents have no chance to escape such conflicts.

Even worse, terrorism—or what's considered as such—has risen dramatically since the end of the Cold War, especially in the 2000s. Up to mid-2013, nearly a hundred terrorist attacks in which at least fifty people died had taken place in the new millennium. The largest of course happened on September 11, 2001, in New York City; Washington, DC; and Shanksville, Pennsylvania, resulting in the deaths of nearly three thousand people.[26]

But add up the death tolls from official conflicts and terrorist attacks, and the grand total still pales in pure numerical comparisons with human history. It's tough to say whether people in developed countries are better off psychologically, though, because conflict and the death it brings are harder to see coming and therefore more stress-inducing.

The situation looks very different in the least developed countries, particularly in Africa and parts of the Middle East. According to Sweden's Uppsala University Conflict Data Program, the number of worldwide conflicts has taken a sharp upturn since 2008 largely because of unrest in those regions. The Arab Spring uprisings in countries such as Tunisia and Egypt, as well as ongoing conflicts in places such as Yemen and Sudan, resulted in only one new peace agreement—signed in 2011, the lowest annual number since 1987.[27]

Fortunately, the total number of worldwide conflicts—thirty-seven in 2011—still fell far below the peak of fifty-three in the early 1990s. That trend is encouraging. For a good portion of the globe, the end of the Cold War caused many long-simmering international and intercultural tensions to devolve into open conflict. Once settled, increased economic prosperity, peace, and stability generally followed. It looks like the same is happening in several of the Arab Spring countries, where despotic governments have had to adopt reforms. Technology played a key role in those uprisings, where dissenters organized and reported a great deal of information via social media. Calling the Iranian protests in 2009 and 2010 the Twitter Revolution might be overstepping it, but the underlying point remains valid nevertheless.

It would be naïve to suggest that humanity is getting over conflict, despite some of the positive trends. But it's also not wrong to point out that technology is driving increased harmony and peace between countries by accelerating economic development. The media often paints terrorism as a religiously or ethnically motivated phenomenon, but the majority of people who take part in it usually come from the lower rungs of the socioeconomic ladder. Rising social inequality could therefore lead to a short-term increase in terrorism, but reforms and improving prosperity in general point toward a more positive future. Leaders such as Osama bin Laden are or have been well heeled and ideologically driven, but the rank and file generally aren't. As Charles Stith, former US ambassador to Tanzania, puts it: "People who have hope tend not to be inclined to strap one-hundred pounds of explosives on their bodies and go into a crowd and blow themselves up."[28]

Undeveloped countries, terrorists, and so-called rogue nations such as North Korea therefore represent outliers that have yet to join the steadily improving global prosperity. The rest of the world has found cooperation more productive than fighting, an attitude that will spread to the outliers if and when they are ready, willing, and able.

## ARE BACTERIA SMARTER THAN POLITICIANS?

Even the worst politicians understand the delicate balance they must maintain between enriching themselves and the populace, which means cooperation rather than betrayal also falls into their long-term interests. Some understand this better than others of course. Many critics believe that cronyism is reaching the same heights as during the Industrial Revolution, when robber barons shamelessly exploited workers while living high off their efforts under the enabling and tacit approval of politicians. If that's true, the historical corrective forces discussed in the previous chapter are about to kick in. We're headed either toward revolution or—more likely—wide-scale reforms that will vent some of this inequality pressure. The survival instinct, when pushed, tends to overrule all other instincts, including greed.

Stanford University sociologist Deborah Rogers believes this is likely because inequality doesn't appear to be ingrained:

> *In a demographic simulation . . . we found that, rather than imparting advantages to the group, unequal access to resources is inherently destabilizing and greatly raises the chance of group extinction in stable environments . . . Although dominance hierarchies may have had their origins in primate social behavior, we human primates are not stuck with an evolutionarily determined, survival-of-the-fittest social structure.*

*We cannot assume that because inequality exists, it is somehow beneficial. Equality—or inequality—is a cultural choice.*[29]

Even humble bacteria understand that social consensus rather than biological imperative decides the Prisoner's Dilemma. Chemists found in 2012 that microbes not only "chat" with each other and play a version of the game, but they also often tend toward cooperation rather than betrayal when making key decisions about their colonies. In this way, they pick the optimal choices for each member when it comes to issues such as cell stress, colony density, and the inclinations of neighboring cells. As head researcher José Onuchic puts it:

*Bacteria that previously existed harmlessly on the skin, for instance, may exchange chemical signals and reach a consensus that their numbers are large enough to start an infection. Likewise, bacteria may decide to band together into communities called biofilms that make numerous chronic diseases difficult to treat—urinary tract infections, for instance, cystic fibrosis and endocarditis.*[30]

Assuming that politicians in democratic countries are smarter than bacteria—not always a safe presumption—inequality looks as though it will reach a natural limit before self-correcting mechanisms inevitably take effect.

Despite periodic ebbs and flows of inequality, humanity's economic future is rising. The Occupy movement expressed much of its anger at the 1 percent—households that earn more than $340,000 a year—but the 5 percent, or those with a combined income of $150,000, lies within reach for many families in developed countries. The protest movement claimed to represent the 99 percent, but that wasn't entirely valid. Many people who earn less than $340,000 a year are doing okay in the grand scheme of things. Put another way, the amount of money that we spent on food two hundred years ago accounted for about 80 percent of our annual budget. Today, it's only about 10 percent, meaning that our disposable income has grown dramatically.[31] A good portion of what's upsetting people actually has to do with the hedonic tread-mill, where an individual's happiness level fluctuates in accordance to what other people have, but we'll discuss this in more depth in chapter nine.

Over the past thousand years, the world's population has increased more than twenty-fold while per capita income has grown thirteen-fold. Since the early nineteenth century, however, that trend has reversed, with income growing eight-fold but population increasing only five-fold. Don't take this incredible decrease in global poverty lightly. The world is treading a path toward greater economic equality even if some people are still "more equal" than others. As we've seen, one big problem is on the way to being solved, and another is replacing it. But put into perspective, this new issue isn't as bad. Relative inequality isn't as terrible as the objective misery, suffering, and death associated with real poverty. Our technologically driven economic growth is bringing hundreds of millions of people into a world where they can dream and plan for the future. That's a great development.

It does, however, raise another potential problem. With all these people in developing countries finally getting a shot at having a future, isn't the planet going to explode from overcrowding?

# 3

# Health: The Unbearable Vampireness of Being

*There is only one difference between a long life and a good dinner: that, in the dinner, the sweets come last.*
—ROBERT LOUIS STEVENSON

At seventy-six years, Louisiana has just about the worst life expectancy of any American state. Only Alabama, West Virginia, and Mississippi have it worse. People in the top twelve states typically can expect to live to age eighty, with Hawaiians topping the list of longevity at an average of eighty-three years.[1] Louisiana's substandard showing comes from high poverty and crime and poor education and health care. The opposite generally holds true in the longer-lived states. It doesn't help that Louisiana has to endure one disaster after another. If it's not suffering through some sort of pestilence—such as the Yellow Fever that raged through New Orleans for much of the nineteenth century or the Bubonic Plague that followed in the early twentieth century—then weather calamities like Hurricane Katrina are throttling it. Such is life in the "Big Easy" and its surroundings.

If the prevalence and commonality of death has had any positive side effect on the state, it's that residents have attuned themselves to its context. "Early on, I got some sense of history and how ages compare and how one of the responsibilities we face in this age is to be conscious of what's unique to it, insofar as we can, and make intelligent decisions as to what's available to us," says Anne Rice, one of New Orleans's most famous daughters.

"You can't do that if you know nothing about ages past or if you believe lies about ages past. If you're aware that in 1850 people starved to death in the middle of New Orleans or New York, that's a dramatic difference between past and future. I'm fascinated by it. Why everybody isn't, I don't know, but I am."[2]

Rice's classic novels—*Interview with the Vampire*, *The Vampire Lestat*, *Queen of the Damned*, and many more—predate the current vampire craze. The undead monsters first inked into English by John Polidori, Lord Byron's physician, pervade the culture now unfortunately: schmaltzy movies, teen-angst books, and soft-core porn TV shows. Rice's oeuvre still stands above most of the genre, however, because it represents a unique approach still not replicated even decades after many of the books first appeared. New Orleans, the veritable antithesis of the Anthropocene epoch, framed Rice's perspective as she grew up there. Modern metropolises have transformed their environs into finely tuned systems of order, but the Crescent City teems with a charmingly antiquated natural chaos. Centuries-old buildings in the French Quarter creak and lean while roots of ancient oak trees burst from the sidewalks of the nearby Garden District. The residents prefer it this way, like listening to a vinyl record with its pops and hisses that tell their own story. The city offers a living, breathing reminder of the past and therefore of how far humanity has come.

"The failure of most vampire literature is that the authors can't successfully imagine what it's like to be three hundred years old. I try really hard to get it right," Rice says. "I really love taking Lestat"—her most famous character—"into an all-night drugstore and having him talk about how he remembers in 1789 that not a single product there existed in any form that was available to him as a young man in Paris. He marvels at the affluence and the wealth of the modern world."

To our caveman friend Blarg, modern humans might appear not unlike Lestat and his vampire kin. We don't necessarily consume blood to live, nor can we transform into bats, wolves, or mist, but we do have a host of seemingly superhuman powers, much like vampires. Chief among those, to the primitive human, would be our ability to live long lives.

If Blarg were exceptionally lucky, he might have made it to his forties, but he more than likely would have succumbed to pneumonia, starvation, or injury before his early twenties—if he survived infancy in the first place, that is. Life expectancy for humans more than ten thousand years ago was short and didn't improve much for a long time. In ancient Rome, the average citizen lived to only about age twenty-four. But most counted themselves fortunate to get even that far; more than a third of children died before their first birthday. A thousand years later, expectations looked much the same.[3]

Over the course of the next eight hundred years, people in the more advanced parts of the world added only fifteen years to their life expectancy. An average American in 1820 could expect to see thirty-nine. Lifespans started to pick up in the early nineteenth century—around the same time that vampire myths were proliferating in Europe—and really sped up in the twentieth thanks to a decline in infant mortality but also because of improvements to health in general. By 2010, the average US life expectancy had nearly doubled from two centuries prior, at seventy-eight years, with similar results in other developed countries. Today, people in Japan live longest, averaging about eighty-two years.[4] To a caveman or an average Roman, that would seem like an eternity.

Rice and many others from the Pelican State recognize this perspective. Even with their state's comparatively low life expectancy, they're still far better off than most people at any point in history. "I would be dead if we were in the nineteenth century," says the septuagenarian. "I'm a type-one diabetic; I'd be long dead. I probably would have died three times over from things that have happened to me. But we're living in the most wonderful age, it's just the most incredible age because never before has the world been the way it is for us. There's never been this kind of longevity or good health. It's incredible to have this many people living in harmony and peace with their fellow human beings, having so many choices of how they want to live and where they want to live, what they want to do with their lives."

## THE VAMPIRES OF OKINAWA

As Rice implies, life expectancy ties closely to economic prosperity, which we've seen relates to technological development. Like economic growth, human health has seen similar dramatic improvements throughout history on account of technology. Longevity hasn't quite become exponential, but it has been profound in recent times and looks likely to accelerate even more in the future, in step with economic advancement. We tend not to think about these facts that are vital when assessing whether humanity is better off because of technology.

Countries with the biggest improvements participated in the Industrial Revolution early and saw better food production and fewer famines as well as stronger resistance to and treatment of diseases. As the Organization for Economic Cooperation and Development puts it, "Increases in life expectation are an important manifestation of improvement in human welfare . . . There has been significant congruence, over time and between regions, in the patterns of improvement in per capita income and life expectation."[5]

Of course, this tremendous explosion hasn't happened everywhere. Life expectancy in many African nations still hovers around forty, and in Swaziland a typical person can expect to make it to just thirty-one. The average for the continent as a whole is fifty-two, whereas Western Europe clocks in at seventy-eight.[6] On the plus side, the projections for less developed parts of the world, just as with their economic prospects, are pointing in the right direction. Crude birth rates—the number of people born each year per thousand members of the existing population—are plummeting. In 1900, forty-two people were born in the developing world for every thousand already there; by 2050, that figure will drop by almost two-thirds to fifteen new births.[7] This decline mirrors what has happened in every developed nation: As each country industrializes and its people rely less on agriculture and subsistence lifestyles, their economic fortunes improve, and families can have fewer children because infant mortality declines and they don't need as many hands to support the household. In more advanced economies, couples focus more on their careers and providing greater resources to the fewer children they do have.[8]

The unfeeling science of demographics does mask the unfortunate truth that life expectancy estimates often skew downward because of the large numbers of children who die early. Again, the trend here also looks positive: Infant mortality is declining dramatically in many developing nations. In some African countries, it's falling by as much as 8 percent a year. The continent overall is experiencing a faster decline than anywhere at any point in history. Like the largely untold news about decreasing poverty in the developing world, this is "a tremendous success story that has only barely been recognized," according to the World Bank.[9] The life expectancy gap between advanced and developing nations is narrowing. In 1950, people in more developed regions could expect to live to around sixty-six, compared to forty-two in less advanced countries. By the end of 2015, that twenty-two-year gap will have declined by half to eleven years.[10] Slowly but surely, the world is equalizing in this measure as well.

On the other side of the equation, life expectancy is going up because people aren't dying as early as they did, and that has almost everything to do with science and technology. Dietitians and food scientists continually uncover better ways to eat, like the traditional Okinawa diet. Even among their already long-lived Japanese compatriots, Okinawans stand out. Living on the group of islands rather than on mainland Japan makes a person five times more likely to hit the century mark because of a typical food regimen low on meat but rich in nutrients and calories.

Scientists studying this diet have found that it significantly lowers the risk of heart disease and several of the most prevalent forms of cancer.[11] Nutritionists and food producers around the world are absorbing those lessons and passing them on. Indeed, Trader Joe's stocks seaweed snacks —coming soon to a table near you, seaweed sandwiches?

Conversely, the obesity epidemic is growing, and some demographers believe it could seriously affect life expectancy in many countries over the long haul. So far, it's a relatively recent phenomenon that has yet to show up in overall numbers. Simple economic growth may end up countering at least some of the problem since plenty of evidence shows that more affluent people tend to eat better or healthier. They stop going to McDonald's and start eating more wholesome, organic foods. A 2013 report from the Center for Disease Control (CDC), for example, found that "as income increased, the percentage of calories from fast food decreased."[12] The other half of obesity comes from a sedentary lifestyle, a trend that may worsen as automation and robots replace more forms of physical labor. One of the biggest challenges facing today's governments lies in countering this natural trend toward inertia and pushing for physical education.

Still, many of the top health-related causes of death in the world are declining steeply. Tuberculosis, for example, ranked among the top ten killers in the world in 2008, yet it had fallen off the list completely by 2011. In the United States, its incidence has plummeted since 1900. Developing countries, where the disease prevails now, will follow similar trajectories as their economies improve.[13] Treatment of HIV/AIDS, also on the World Health Organization's top killer list, is also improving by leaps and bounds thanks to better drugs and education. The total number of people infected quadrupled from 1990 to 2010, but the overall growth of the epidemic has stabilized because of better anti-retroviral therapies. Overall new annual infections and AIDS-related deaths are declining.[14] As a 2010 report put it, people in their mid-thirties living in developed countries who discover they are HIV-positive today can expect to live another thirty years.[15] Even better news is the growing number of people apparently cured of the disease through bone-marrow stem cell transplants.[16] The governments of advanced countries now need to figure out how to facilitate such treatments for developing nations, where they're needed most.

Even certain forms of cancer are not necessarily the death sentences they used to be. Survival times for non-Hodgkin's lymphoma, breast, and colon cancers have improved dramatically over the past forty years, to the point where scientists are saying they are at "an amazing watershed" in understanding cancer in general.[17] There's still much more work to do with other types of cancer, but the overall trajectory is pointing in the right direction.

## MALTHUS VERSUS MOORE

The product of all these improvements—longer life expectancy—freaks out the Malthusians. The good reverend Thomas Malthus wrote in 1798 that exponential population growth would outstrip the world's food production capabilities, and ever since then people have been predicting shortages resulting in famine, pestilence, and war. It hasn't happened because improvements in food production technology have also turned out to be exponential rather than linear.

World War II in particular ushered in a new era of mass food production as techniques and technologies developed to feed troops overseas came to benefit the general population. Freeze-dried foods, microwave ovens, refrigerators, plastic packaging, and other wartime innovations helped stabilize and boost production and distribution. Fast-food chains such as McDonald's—despite all the negativity they attract—also contributed by standardizing systems and instituting stringent quality testing. If not for them, mass outbreaks of food poisoning likely would happen far more often. The Green Revolution of the 1960s and 1970s also helped: New chemical fertilizers and crop-breeding techniques dramatically increased production capabilities in countries such as Mexico, India, and the Philippines. Even space exploration has translated into better earthly food. Many low-sodium products owe their pedigrees to NASA because astronauts have to watch their salt intake in space. That humble tub of margarine in the fridge just happens to be an indirect byproduct of rocket fuel.

Today, food production is undergoing another technological boom with genetic engineering. It started in 1994 with the Flavr Savr tomato, designed to ripen faster, but the field has expanded to include insect-repelling corn and vitamin-rich rice. Many people fiercely resisted these genetically modified foods at first, especially in Europe, but acceptance among regulators and the public is growing as predicted doomsday scenarios fail to materialize. GM foods, as they are known collectively, hold the promise of applying technological stacking to food production because altered crops can have not just one but several new traits. SmartStax corn, created by chemical company Monsanto in 2009, offers a good example as it resists both insects and herbicides. Future versions could also require less water or sunlight. Like humans, food is gaining technological superpowers. Last but not least, the converse is also happening, where organic foods are becoming more popular. At the moment, they are typically more expensive than heavily processed foods, but with higher prosperity levels allowing more access to them, economics of scale will soon start to kick in. Organic food, in the very near future, will be more affordable and therefore more widespread.

The world's population hit seven billion in 2011. Another two billion will come along by 2050, for a total of nine billion. Experts believe this rate, combined with better economic conditions in developing nations, means current food production will need a boost of nearly three-quarters of its current capability.[18] That's a tall order, but production has kept pace so far and stands poised to continue to do so. Moore's Law is trumping Malthus's dread.

That figure of nine billion doesn't include just new people being born. It also includes people living longer. Case in point: about 40 percent of the girls born in Great Britain in 2013 will live to age one hundred, with boys not far behind. By 2060, that proportion will increase to 60 percent.[19]

In 2003, the UN released a bold report, "World Population to 2300," which sought to predict the next three centuries of global demographics. The authors stressed that it was difficult to guess past 2050, but they estimated that world population would peak at 9.2 billion in 2075, then decline after that.[20] Why? Life expectancy is climbing, and people are dying less frequently, but crude birth rates are falling considerably faster. Families in many countries already are falling short of the population "replacement level" of 2.1 children each. America hit that exact number in 2010, while many other developed nations—including France, Germany, the United Kingdom, and Canada—already had fallen below it. Families aren't making babies fast enough to account for the number of people keeling over. People may be living longer, but everyone dies in the end.

Life expectancy, meanwhile, is "assumed to rise continuously with no upper limit, though at a slowing pace dictated by recent country trends. By 2100, life expectancy is expected to vary across countries from sixty-six to ninety-seven years, and by 2300 from eighty-seven to 106 years."[21] Given the trajectory of the past two centuries, those estimates are conservative, which the report acknowledges. "The diffusion of knowledge and technology, which could narrow the gaps between countries, was not factored into the projection methodology. Such an implicit assumption of independent trends does not affect short-term projections, but it seems to affect long-range projections like this one."[22]

Technology will affect the gaps and also extend life overall. Many scientists believe we're on the cusp of discoveries that will add significant years to average expectancy. Harvard geneticist David Sinclair, who is researching resveratrol—an anti-aging compound found in red wine—thinks 150 years will be achievable soon. "We are going through a revolution," he says. "We might have our first handle on the molecules that can improve health."

Futurists, or those social scientists who try to predict the future, are taking these advances to heart. Singularity prophet Ray Kurzweil, among others, believes that effective immortality lies around the corner. At the rate science is going, he figures that life expectancy will gain a year every year by the late 2020s, which means that if we can make it till then we'll have a good shot at living forever.[23] Kurzweil goes further in his predictions by arguing that the human brain eventually will be reverse engineered, which then will allow for the reconstruction and effective cloning of humans and their personalities. When that happens, we'll be able to transfer our consciousness into machines and virtual worlds and live forever like the Cylons of *Battlestar Galactica*.

Kurzweil's is a controversial view. Many neuroscientists believe the brain, personality, and consciousness may never be understood fully. On the other hand, Kurzweil seems right in suggesting that many of his critics can't see the forest for the trees; specialists entrenched in their own fields often can't take into account advances in other areas as we saw with the Human Genome Project. His belief about replicated immortality may get a shot in the arm with the process of mapping the brain. In 2013, President Obama launched Brain Research through Advancing Innovative Neurotechnologies—yes, BRAIN—while the European Union is funding the similar Human Brain Project. Doubts will arise about what results such efforts will net, but they likely will be important and happen more quickly than some observers expect.

Think of some of the past mind-blowing breakthroughs. In 2012, researchers at Wake Forest University and the University of Southern California developed "memory chips" to turn specific memories off and on in a mouse's brain. An implanted mouse might remember how to navigate a maze perfectly one minute, but with the simple flick of a switch that same rodent instantly forgot everything it knew.[24] Even more impressive was a successful experiment by UC Berkeley scientists to create videos from memories using an fMRI machine. After being exposed to random YouTube videos, a computer reconstructed test subjects' visual recollections with a fair degree of accuracy.[25] It's the science fiction of *Total Recall* and *Inception*, and it's already happening.

The near-term reality amid all these life-expectancy projections probably lies somewhere in the middle. Human longevity will stretch longer than the UN's technologically absolved prognostications and possibly improve significantly through coming advances. Either way, humans are living dramatically longer than even our forebears of only a century ago. This new reality is having and will have major repercussions not just on society and our cultural institutions but on who we are as people.

## NETFLIX FOR THE HEART

One question that inevitably arises when talking about living longer is, are we living better? The last chapter answered that question on an economic level, but what about health-wise? A person might live to a hundred today, but what's the quality of those latter years? Is it worth it if that means living in pain or with severe physical limitations? As Ciaran Devane of Macmillan Cancer Support puts it:

> *While it is wonderful news that more cancer patients are living longer overall, we also know they are not necessarily living well. Cancer treatment is the toughest fight many will ever face, and patients are often left with long-term health and emotional problems long after their treatment has ended. For instance, of those colorectal cancer patients still alive between five and seven years after their diagnosis, two thirds will have an ongoing health problem.[26]*

At the very least, more of us will look forward to a cocktail of pills every day during our twilight years. Is that reality, where we all become drug dependents, worth it? Kurzweil thinks so. He admits to taking 150 different pills—mainly vitamins, supplements, and preventative medications designed to slow down aging—so he can make it to the Singularity. Quality of life is purely subjective, he says, until you come face to face with mortality. "I've spoken to many 100-year-olds, and if you ask them if they want to live to a hundred and one they will tell you they do."[27]

The question of living better can't be answered empirically unless we consider what used to make us sick and kill us. The top three killers of Americans in 1900—pneumonia or influenza, tuberculosis, and gastrointestinal infections—don't appear on the 2010 list, banished to manageability along with historic illnesses such as smallpox, scurvy, and rubella.[28] Today's top three—heart disease, cancer, and noninfectious airways disease—stand apart. Unlike their predecessors, they're not infectious; instead they're environmental, self-inflicted, or genetic. Some doctors believe that makes them eminently more treatable, perhaps with the sort of preventative self-care practiced by the likes of Kurzweil. Others think we're entering a technology-driven health care revolution that not only will beat back some of the worst killers but also greatly improve the quality of life after illness.

Cardiologist Eric Topol researches genomics at Scripps Research Institute in La Jolla, California, and speaks regularly at events such as the annual Consumer Electronics Show (CES) in Las Vegas, Nevada; the Technology, Entertainment, and Design (TED) talks; and *Wired* magazine conferences. He often presents the same message that he did in his 2012 book, *The Creative Destruction of Medicine*: The health care industry is undergoing a bottom-up transformation thanks to digital technology that will result in longer and better life.

Topol hasn't used a stethoscope in years. Instead, he uses a relatively inexpensive handheld ultrasound device, which preempts the need for an electrocardiograph and lets him share results with patients right away. Nor does he prescribe a Holter monitor, a complicated and uncomfortable heart-monitoring contraption that patients need to wear for long periods of time. Instead, he gives them a much smaller and cheaper monitor that sticks to their skin, much like a bandage, that patients then mail back to his office. The monitor, he says, is like "the Netflix of heart-rate monitoring."[29] He also applauds the heart-monitoring apps released for smartphones and tablets. He makes a point of asking new patients what phones they're using. If he doesn't like what he sees, he gives them some unusual advice. "The prescription I give them is, 'You need to get a new phone.'"

Gadgets and apps like these are proliferating quickly and comprehensively. With inexpensive heart-rate monitoring and step-counting wristbands such as the Nike Fuelband and the Fitbit becoming a hit with consumers in recent years, inventors and entrepreneurs are flooding the market with all manner of self-tracking tools. Such devices—from the June, a stylish bracelet for women that monitors ultraviolet light exposure, to the Muse, a thin headband that monitors brainwave activity and advises on relaxation techniques—are overrunning the CES. Sensoria's Fitness Socks gauge how well you walk, your individual footfalls and gait, then give you an overall sense of your foot health. Kolibree's connected toothbrush tracks how well you brush your teeth, meaning that you don't have to wait until the dentist's office to know how poorly you're doing it. The HapiFork tracks how quickly you eat and chew and buzzes if you're going too fast. No human function or activity can't be tracked, measured, and corrected. As Moore's Law kicks in, all of these gizmos and apps will get better and cheaper.

The proliferation of self-tracking means doctors and individuals are assembling an increasing wealth of data, which inevitably will cause health care to become more personalized. That development runs counter to the population-based method of medicine administration, where pharmaceutical companies and doctors effectively guess based on sample groups. For much of the modern era, pharmaceutical companies have manufactured drugs after testing them on a relatively small test group of people, many of whom don't include or represent the final patients. Like fingerprints, every person is different because he or she possesses a unique biological makeup, complete with its own nuances and combinations of health conditions. That's why so many mass-market drugs either don't work or come with a terrifyingly long list of possible side effects. With better data and the spread of individualized health information, pharmaceutical companies increasingly can specialize and improve drugs and treatments for smaller groups of people, as they have with certain types of cancer and cystic fibrosis.[30] Topol thinks diabetes, for example, can break into twenty-five different subtypes, with different drugs for each. Such medicine will be more effective and carry fewer side effects.

On the diagnosis side, supercomputer assistants—some derived from the likes of IBM's *Jeopardy* champion, Watson—are aiding doctors in crunching all that data to generate better assessments.[31] Put all those pieces together, and those longer lives that people are experiencing don't have to teem with pain and misery. "People don't want to make it to a certain year, they want to make it to a certain quality of life," says Topol. "Decreasing the burden of chronic diseases, that's where it's at. This will transcend the old dinosaur era of medicine."

The term "moonshot"—originally coined to describe NASA's Apollo program that put Neil Armstrong and Buzz Aldrin on the moon—over the past few years has become a metaphor to describe undertakings by companies and institutions to solve humanity's huge issues. In 2013, Google announced one such moonshot, the launch of the California Life Company, Calico for short, aimed at extending life and improving its general quality. It's an unusual move for a company that makes most of its money from Internet advertising, but Google's founders insist that they have a responsibility to use their wealth to improve humanity. But Calico's researchers will be looking at health concerns from a different perspective. Google cofounder and chief executive Larry Page puts it this way:

> *Are people really focused on the right things? One of the things I thought was amazing is that if you solve cancer, you'd add about three years to people's average life expectancy. We think of solving cancer as this huge thing that'll totally change the world. But when you really take a step back and look at it, yeah, there are many, many tragic cases of cancer, and it's very, very sad, but in the aggregate, it's not as big an advance as you might think.*[32]

As with just about everything Google does, Calico may have entirely unexpected results. Who knew that when Google first launched its search engine in 1997 that it ultimately would become an advertising behemoth? And who knew that Google Maps would eventually spawn Street View, a tool with which we can virtually view the world?

Whether we're living better is one of the most subjective questions we can ask ourselves since so many factors come into play. Age and era are the biggest. If you had asked a thirty-year-old in the eighteenth century to rate her quality of life, her answer would have differed widely from a similarly aged person living in the developed world right now. Yet a ninety-year-old today might feel the same as that thirty-year-old three centuries prior. Neither person would have the necessary context to consider the other's life.

## LIFE AS A COMMODITY

"We're seeing death in a new way," says Rice. "Instead of taking it for granted, the people I know see it as a personal catastrophe. I get e-mails from people who are actually surprised that someone has died. They regard it as an injustice. I understand their feelings,

I get it, but this is a fairly new perspective on death. Nobody in the 1900s would have regarded death as a personal catastrophe. They would have mourned and might have been grief stricken, but they saw death all around them."

In that sense, death as an event is increasing in its importance, which conversely means that the value of human life is also rising. Imagine how valuable human life will become if we do manage to clone ourselves digitally as futurists believe. How tragic would the loss of a person be then, where perhaps they are wiped out by some sort of computer virus when their entire being could otherwise be preserved forever?

In economics, a commodity is more valuable the rarer it is, which is why a finite resource such as oil can fetch top dollar. Human life, measured as time on Earth, works in the same way— but it also doesn't. If we can expect to live a long time, we may not treasure individual years as much. We might even waste some by indulging in extraneous pursuits, such as sailing around the world or mastering the ukulele. On the other hand, if we live for many years, the value we have to other people, such as friends and family members, tends to increase.

Life differs most from commodities in the value it has for the person possessing it. If an individual has some oil but doesn't like it, he or she can sell it for a nice profit because other people or entities do value it. An individual's life, however, doesn't have the same transferable value. Relatives and loved ones may treasure your life, but ultimately it isn't worth much if you don't yourself, which is where quality comes in.

A host of factors determines quality of life, beginning with a biological baseline. As the old cliché goes, you don't have anything if you don't have your health. But obviously what we do with our lives, the relationships we have, and the degree to which we meet our goals and dreams, determines the value we place on them.

At the beginning of *Interview with the Vampire*, Lestat is a confident and happy undead monster in late-nineteenth-century

New Orleans. He creates a vampire family of sorts by siring an aristocrat named Louis and then a young girl named Claudia. They live together happily for a while, but eventually his protégés turn on him and flee. Near the end of the book, in modern times, Lestat is living in squalor, barely alive. Despite his immortality and the automatic fulfillment of his biological baseline—as long as he drinks blood—he's miserable after years of being alone with the memory of his family's rejection. The message is clear: It's not enough simply to exist. Other necessary factors bring value to life.

Most vampire fiction falls firmly into the realm of horror, but Rice's books read more like psychological case studies; the vampire characters comment on the human effects of technology and progress or project themes or ideas discussed in a previous chapter. With technology extending our lives and improving our health, the analogy fits better today than ever before: Humans may not be vicious psychopaths who drink the blood of innocents, but we are becoming more akin to vampires in that way. Age affects Rice's characters in different ways. Some become wise and contented while others grow vain and egotistical. Which path are we treading as we inch toward immortality?

As a trope, vampires represent a subconscious desire for immortality. Some of us think about dying from time to time and even plan for it, but we're not good at imagining what realistically might happen when it does occur. "We can't conceive of our own death," says Rice. "It's just not possible, and yet we're living mortality every moment that we're alive. The more I see people die and experience the death of those I've loved and the more time I spend in sick rooms seeing people die, the more I'm aware that death takes people unawares. There's no way you can prepare for going into another country of existence or winking out as if somebody simply pinched a candle flame out. We like to imagine we're vampires because we feel really comfortable doing that."

# 4

# Jobs: A Million Little Googles

*All of this has happened before, and it will all happen again.*
—J. M. BARRIE, *PETER PAN*

No one likes cleaning the toilet, but unfortunately, advancements in the technology to do so automatically have been slow in coming. iRobot—the company that makes the Roomba automated vacuum—released the Scooba floor cleaner in 2011, but it's too big to squeeze behind a toilet, never mind the bowl itself. The field clearly still has a long way to go.

When coming up with new types of jobs for robots to do, creators often focus on the three Ds: dull, dirty, and dangerous. Cleaning the toilet hits at least two of those, which is why someone, somewhere, is putting a great deal of thought and effort into the situation; we just haven't heard about it yet. Or, at least, we should hope somebody is working on it. A robot toilet cleaner is inevitable, and I'll be first in line when it does come out. Moore's Law, do your thing.

Machines have been doing what we don't want or can't do for as long as humans have walked the earth. A machine, by its simplest definition, absorbs energy of one sort and transforms it into a more useful form. Blarg the caveman benefited from a form of mechanization when he used his wooden club to beat dinner into submission: Force + Mass = Yum! Since then, machines have made it easier to craft pots and plates, harvest crops, traverse distances, lift heavy objects, cool the air, open cans, direct traffic, go up and down floors in buildings, wash dishes, dry hair, brush teeth, wax floors, make coffee, and on and on. Machines also have made possible the seemingly impossible: printing huge volumes of books, flying through the air, walking on the moon, and peering inside bodies without cutting them open.

Technology has radically and, yes, exponentially redefined what we humans do and can do. Only two hundred years ago, nearly three-quarters of Americans, for example, worked on a farm. Machines have all but replaced people there, with only about 1 percent of the workforce still engaged in farming.[1] As mechanization arrived, farmers and farmhands moved to factory jobs, then to office work. In each case, the majority had no idea what the future held in store. Only a prescient few farmers might have seen that their jobs were becoming obsolete and that they would have to learn how to operate an industrial lathe or machine press. In turn, a small minority of factory workers anticipated that they eventually would need to wear a shirt and tie and work at a desk with a computer.

The same is true today except—as we well know by now— our machines are becoming much better really quickly. In recent decades, they've acquired virtual brains, which means yesterday's simple automatons are now capable of doing much more. The humble thermostat, for example, can tell when you're not home and lower the temperature accordingly. An automated investment algorithm can buy and sell stocks without your ever knowing. Your computer or phone routinely suggests what you might like to see, do, or purchase. Anyone keeping a list of what machines and robots can't do—and, in many cases, do better than humans—is presiding over a quickly dwindling list.

What we actually do is undergoing a similar process of upheaval. The jobs of the Office Epoch are rapidly yielding to automation and robots, just like machines made the Farming Epoch of yesterday obsolete. Some estimates figure that by the end of this century, machines will have replaced humans in nearly three-quarters of the jobs we're doing now.[2] This trend doesn't apply just to human factory workers, who some say are largely obsolete already. It covers pretty much any task that involves repetition or that doesn't include spontaneous and creative thought.

In early 2013, a robot doctor started making the rounds at Daisy Hill Hospital in Newry, Northern Ireland. Controlled by a human more than twenty miles away, the machine enables doctors to teleconference with bed-ridden patients. While a human doctor is still in the loop, the automaton—which resembles one of the terrifying Daleks from *Doctor Who*—obviates the need to have an actual physician there, which means one less job. The Federal Drug Administration began approving such machines the same year. The da Vinci "robot surgeon," meanwhile, has been in use since 2000, assisting doctors with operations ranging from hysterectomies to hernia repair. The machine doesn't perform surgery autonomously yet, but its software is capable of doing so. It's easy to imagine a sophisticated, rock-steady robot doing this kind of work better than a human prone to nervous, shaky hands.

In Harbin, China, an entirely mechanized restaurant, complete with robotic cooks and waiters, opened in 2012. In Beijing, chef Cui Runquan debuted his Noodlebot the same year. The humanoid machine slices noodles much like a car's window wipers operate, with a constant back-and-forth motion impossible for a human to maintain for long periods of time. "It is the trend that robots will replace men in factories," Runquan says. "It is certainly going to happen in sliced-noodle restaurants."[3]

For its part, Google is working hard to make self-driving cars a reality. Riding in one at the Consumer Electronics Show in 2008 and watching the steering wheel turn itself as the vehicle careened around the test course was one of the most sublime experiences of my life. But this isn't just another of the company's pie-in-the-sky experiments; automakers have jumped on the bandwagon. Audi and Toyota are working on autonomous cars, with Nissan pledging to bring "multiple affordable, energy-efficient, fully autonomous-driving vehicles to the market by 2020."[4] A lot will change when this happens. Taxi drivers, for one, will have to look for new lines of work since we'll be able to order up an automated cab with our smartphones. Even the idea of car ownership may be thrown into flux too. Who will want the trouble and expense of owning a vehicle when you can have one pick you up and take you wherever you need to go with a few taps on your phone? If we don't own cars, we won't need to pay insurance on them, which will affect the insurance industry and the people in it who raise our rates for a living.

At the Yotel in the Hell's Kitchen neighborhood of New York City, a machine greets guests. There's no front desk but rather a bunch of self-serve kiosks, like those popping up in supermarkets, drugstores, and other retailers. The machine takes your relevant information, including credit card details, spits out a room key, and tells you to have a nice day. At the other side of the lobby, a giant robot arm—the kind used in car-manufacturing plants—lifts and places luggage into a cupboard-like storage unit for people who have checked out but need to store their belongings. In my case, the check-in and check-out process was significantly more efficient and pleasant than a regular hotel, which is undoubtedly unnerving front-desk employees in the know.

In Atlanta, the Monsieur robot is putting a similar fear of God into bartenders with its ability to make any one of some three hundred cocktails, and in just seconds. At the pharmacy at the University of California at San Francisco, a computer receives prescription orders, while robots package and dispense them. South Korea is planning to open a fully automated robot theme park, complete with rides, by 2016, which may force Korean carnies to the unemployment line. Even those few people who still work on mega-farms might be displaced further. In France, wheeled robots are pulling weeds from around lettuce and pruning vines in vineyards.

Virtually no one is safe—not even us book writers. The San Diego–based Icon Group International already has hundreds of thousands of robot-written books to its credit for sale on Amazon.

Most aren't any good, but . . . there's Moore's Law again. Someday, this book and others like it, which seek to summarize the effects of technology on people, will probably be written—ironically—by robots.

Perhaps we all should get jobs cleaning toilets since that field still seems wide open to humans. We could end up like the once-mighty horse: Just two centuries ago, horses powered everything from labor to transportation, with one horse or mule for every three people. Today, that ratio has dropped to one for every ten people simply because we don't need as many of them anymore.[5] Now most horses get to lounge around and eat hay and perhaps go for a run a few times a day. (Come to think of it, being a horse might not be so bad.)

The rise of robots and automation is understandably causing a good deal of angst. British economist John Maynard Keynes summarized it well back in 1930 when he spoke of the "new disease" of technological unemployment. New advances at the time—such as better methods of steel production, the spread of the automobile and frozen foods—hinted at the "means of economizing the use of labor outrunning the pace at which we can find new uses for labor."[6] People have worried about automation and robots taking their jobs for the better part of the past century, but with technology advancing so quickly, those fears may well be coming true. Like our farming forebears, we have no way of knowing what's going to happen next.

Or do we?

## JUSTIN BIEBER AND POLKA

With plentiful data from several sources that generally agree with one another, exponential change in employment is relatively easy to measure. The key figure is productivity or the average measure of the efficiency of production. Economists use that number to determine how much humans must work to produce one widget or unit of service. In the first decade of this new millennium, US productivity growth reached its highest level since the

1960s—nothing short of miraculous—with similar increases happening in Europe.[7] Rapid advances in technology easily explain the "miracle," however. Productivity, or how much one worker can output, generally increases in one of two ways: Either more workers are hired so that each individual's workload goes down, or new technology makes that person able to do more.

Additional workers certainly can't explain the miracle. America experienced zero net job creation in the aughts.[8] The Great Recession ended with what many economists described as a "jobless recovery." The recovery itself was easy to identify because of the presence of one of its sure signs: Businesses once again started buying computers and other technological equipment. Such spending almost always dwindles during a recession. The rebound was harder to proclaim, given the lack of another sure sign: the resumption of hiring. In 2013 the unemployment rate was still higher than in 2007, but other signs pointed toward recovery.[9] Companies had enough confidence to spend on new equipment but not on new people. What was happening?

As Sherlock Holmes would say, once the impossible has been eliminated, whatever remains must be the truth. In this case, businesses were thinking they could produce the same or more widgets by adding machines rather than people. Suddenly, Keynes's warning from a century ago looks prophetic. In *Race Against the Machine*, economist Erik Brynjolfsson and MIT research scientist Andrew McAfee raise precisely that specter: The rate at which machines are displacing human jobs has outstripped the pace at which we're creating new jobs. The transitions from farming to factory work to office employment resulted in veritable tsunamis of new and better jobs, as did the early stages of the digital revolution. But the accelerating wave of technological growth means we aren't thinking up new, better jobs fast enough. The terrifying alarm is sounding for everyone from humble office workers to the highest policy makers. What should we do in this encroaching age of human obsolescence?

It's actually not as bad as it sounds, since the solution is relatively straightforward. One of the side effects of fast technological advancement is the even faster growth of combinatorial development, which all that exponentialism makes possible. Remember the technological "stacking" we saw in chapter one. Virtually every new product or service available today resulted from numerous prior developments working in conjunction with one another.

Just as rap and hip hop take beats, riffs, and other samples from existing songs and mix them with new sounds and vocals, so too does technology. YouTube, for example, isn't just a place where people post mash-up music videos—say, cats dancing to Justin Bieber vocals over polka beats. The site itself is a mash-up. It exists because its developers took advances in broadband network capabilities, video viewing technology, and user interfaces—to name but a few—and combined them into something new and useful. Every new product and service is a variation on this theme. As technological advancement continually makes new goods and services possible, the possibilities become infinite. "Combinatorial explosion is one of the few mathematical functions that outgrows an exponential trend," write Brynjolfsson and McAfee. "That means that combinatorial innovation is the best way for human ingenuity to stay in the race with Moore's Law."[10]

The angst over what humans will be doing in the future, therefore, isn't coming from a lack of possibilities but rather from our inability to imagine them. It's an amazing contradiction of human nature: We're incredibly imaginative, but we can be remarkably short-sighted especially when it comes to thinking about our own future. As with food production, we can be remarkably Malthusian in our thinking when it comes to jobs. Stanford University economics professor Paul Romer sums it up succinctly: "Every generation has underestimated the potential for finding new recipes and ideas. We consistently fail to grasp how many ideas remain to be discovered. The difficulty is the same one we have with compounding. Possibilities do not add up. They multiply."

## CHRISTMAS IN ISRAEL

Jonathan Medved is a master multiplier, but he looks a little like Santa Claus—or at least what jolly Saint Nick might resemble on the other 364 days of the year. Decked out in a colorful Hawaiian shirt and leaning back in his chair, hands folded over his slight belly, Medved mirthfully explains the success of Israel's high-tech sector, a hotbed of exactly the sort of multiplying opportunity that Romer means. Medved finds a good deal of irony in it, given that "President Whackjob wants to wipe us off his map." He's talking about Iran's Mahmoud Ahmadinejad, but others no doubt will follow.

"This place is risk central. We live risk, we eat risk. That's real risk, and I worry about that," says Medved, punctuating the point with a hearty chuckle. "The risk of starting a company and losing my job or somebody some money, that just doesn't compute. I'm not cavalier about other people's money or jobs, but here that doesn't qualify as risk."[11] (Two weeks after my visit, the Tal Hotel, where we talked, was closed down because of missile attacks from Gaza.)

Medved is only in his forties, so, despite his graying beard, he's not the right age to play Santa Claus. But as one of Israel's top venture capitalists, he's certainly in the right line of work. Like many of those in whom he invests, he is himself a serial entrepreneur. He has helped sell startups to eBay, Alcatel, Corning, and VMWare, and he launched a number of companies onto the NASDAQ stock market, including Compugen and Accent Software. He cofounded Vringo, a venture capital company specializing in mobile social and video applications, as well as Israel Seed Partners. His latest business is OurCrowd, a crowd-funding organization for angel and accredited investors. When someone has a great idea, they see Jon Medved. If he likes it, he makes Christmas—or Hanukkah if you like—happen by dispensing the money.

You can drive from one end of Israel to the other in five hours. It's smaller than New Jersey, and, at just about eight million people, it has a population smaller than New York City's. Yet, as Medved stresses, Israel is a giant in the world of technology. The numbers prove it. Only two nations—America and China—have more companies listed on the NASDAQ. With one startup for every 1,844 people, Israel has the highest density of such companies in the world. Per capita venture capital investment, meanwhile, is 2.5 times greater than in the United States, thirty times that of Europe, and eighty times more than China. In absolute dollars, Israeli companies attract as much funding as their UK counterparts or France and Germany combined. As in Silicon Valley, this forward motion is happening despite global economic fluctuations or—in a situation largely unique to Israel—frequent armed conflicts. The country's share of the global venture capital investment market doubled during the first few years of the new millennium, despite the bursting of the tech bubble, a war with Lebanon, and ongoing clashes with Palestinian groups.[12]

The result is a country awash in entrepreneurs. Waitresses, bar patrons, shop clerks, and even street performers are either involved in a startup or know someone who is. The startup wave, seeded and nurtured by government programs decades ago, has also helped attract big multinationals. With so many tech-savvy, risk-hardened, entrepreneurially minded citizens, the country represents a gold mine of talented employees for the likes of IBM, Intel, Microsoft, Google, Apple, Motorola, and others, all of which have set up research-and-development centers in the country. Despite all this, few people know how much of consumer technology is designed in Israel, from the processors in our computers to the websites we use to shop.

All of this innovation has had a stunning effect on the country's economy. When established in 1948, Israel had a standard of living comparable to America in the 1800s.[13] But with its strong push into the high-tech sector over the second half of the twentieth century—helped of course by strong US aid—it experienced sixty-fold economic growth in just five decades. Now Israel ranks among the most highly advanced economies in the United Nations' Human Development Index.

At the core of its success lie humble startups and the plucky entrepreneurs behind them. They come from all walks of life, ages, and professions. Musician Yoni Bloch, who felt that music videos weren't interactive enough, started Interlude, a sort of online choose-your-own adventure video maker. Michael Schulhof, former president of Sony America, co-founded Anyclip, a Google-like contextual search engine for video. Parko, a crowd-sourced app that helps users find parking spaces, is the brainchild of the young Itai David, ironically driven to our meeting by his father.

Most of the startups are tiny, with only a handful of people working on them. Sometimes they consist of just a single individual. The majority will fail; some will be bought; a few may break out and become big businesses. Few are seeking or getting big venture capital money because most of the tools they need are cheap and available online. Many are receiving investments to the tune of one hundred thousand dollars rather than the millions they might have required years ago. It's another testament to the price-performance improvements both to hardware and software that this is happening on a business level.

Given the small domestic potential and generally hostile conditions, virtually every startup is focused on the global market. Israeli kids tune in to the same Western zeitgeist touchstones as any teenager growing up in Idaho or Winnipeg. They watch *Breaking Bad*, listen to Jay-Z, and line up for the latest iPhone. Put all that together, and it's often impossible to identify Israeli startups as such: Abesmarket.com, the Waze app (purchased by Google in 2013 for a billion dollars) or Microsoft's Kinect look like they could have been created by Westerners for Westerners.

Whatever the fate of each individual startup, their founders are more than likely—whether they succeed or fail—to go back to the well again. Some who flame out may go work for a big multinational until they've restocked their savings account, then try again. Medved guesses there are five or six thousand serial entrepreneurs in Tel Aviv alone. Many have had duds, but plenty have also had a hit or two, which makes it easier to trust them with money. Dov Moran started M-Systems in 1989. The company invented the USB flash drive and was ultimately acquired by Sandisk for $1.5 billion. Moran then went and started Modu, an ambitious company that wanted to build the world's smallest mobile phone. "It was a total failure and must have lost hundreds of millions of dollars," Medved says. "But now he's back with a whole bunch of companies and I just wrote a check today to one of them."

It's this acceptance of risk by all parties that defines Israel's economy. The ultimate payoff is that by making things the rest of the world wants, they're also creating jobs. Investors, meanwhile, are pouring money into the country. Sure, they lose some of it, but they wouldn't keep coming back if the return ultimately wasn't there. The big companies are also piling in, hoping to take advantage of the huge talent pool. In Israel, everyone is winning and there are more than enough jobs to go around, with the politicized exception of the marginalized Arab minority. But by and large, no one is worried about robots; the country is ironically one of the world's biggest makers of them, per capita.

For the most part, Israelis are doing exactly what they need to do to stay ahead of the machines. They're focusing on jobs that stress leadership, team building, and creativity or skills that robots don't yet have and likely won't for a while, if ever. Virtually every new startup is harnessing the combinatorial power of technology: Interlude is using all the same elements that YouTube did, plus YouTube itself; Anyclip is incorporating advances in search algorithms with faster broadband speeds and processing capability; Parko is harnessing developments in GPS, motion sensing, and wireless networks. In this ongoing war against the machines and their efforts to make humans irrelevant, Israel is becoming the template for how to fight back.

"The ability to compete becomes tied to the ability to innovate," Medved says. "If you can't be part of the innovation economy, you're going to be a laggard. The good thing about innovation as a resource is that it never runs out."[14]

## GARAGE INC.

But can this successful recipe be applied elsewhere, and can it work in a bigger country? The answers to both questions is "yes," and it's already happening. In the United States, the tech sector is adding jobs three times as fast as the country's economy as a whole, and not just in hubs such as Silicon Valley and Seattle. Nearly every US county—98 percent of them—had at least one high-tech establishment in 2011, with each job in the sector creating more than four jobs in the wider community thanks to significantly higher wages than in other fields. Demand for jobs in the sector, meanwhile, is expected to outpace growth over the rest of US industry in the near-term future.[15] The United States has shed much of its manufacturing capability over the past few decades, but ultimately it's startups that have driven the broader economy overall. Amazon, Apple, Facebook, Google, and their kin—all started in garages and dorm rooms by plucky entrepreneurs—have added thousands of jobs and created entirely new industries.

More importantly, the speed at which these companies grow, like the technology they use, has accelerated dramatically. While blue-chip "bricks-and-mortar" firms such as Procter and Gamble or McDonald's took decades to build, technology-oriented startups are becoming huge in no time at all today thanks to the tools that enable them and the potential of a global market. Both Google and Facebook took fewer than ten years to become stock market giants, and Facebook bought Instagram, the photo sharing service, for $1 billion just eighteen months after the startup launched. Stories like these aren't isolated. They're becoming increasingly common and are inspiring more and more people to become entrepreneurs.

In Canada, the startup revolution is also in full swing. According to a report from the Canadian Imperial Bank of Commerce, half a million people were in the process of starting their own businesses as of June 2011. That's more than ever before, and it looks like just the beginning. As senior economist Benjamin Tal wrote:

> *Irreversible structural forces suggest that the next decade might see the strongest startup activity in the Canadian economy on record. The gradual shift to a strong culture of individualism and self-betterment, the role of technology in driving the transition from boardrooms to basements, the more global and interconnected markets that require greater specialization, flexibility and speed, as well as small-business friendly demographic trends are among those forces that are likely to support a net creation of 150,000 new businesses in Canada in the coming ten years.*[16]

When measuring self-employment and entrepreneurship, it's important to note the difference between necessity and opportunity. In Canada, only a fifth of new businesses began because their proprietors had no other options on the table. The rest quit their existing jobs with an eye either to having a more enjoyable career or to striking it rich as the next Google or Instagram. This high proportion of opportunity-driven entrepreneurship is good. As Tal says, "With more business owners starting operations by choice, their likelihood of success may increase," which is why the country's net success rate is going to rise to a whopping third of businesses. Opportunity-driven entrepreneurship also closely correlates with a country's particular economic situation. Those better off generally have higher numbers of voluntary entrepreneurs, while self-starters in poor countries usually have no other choice. Europeans, for example, are nearly three times more likely to be improvement driven than necessity driven, while in Africa the ratio is almost even.[17] With the economic fortunes of the developing world improving, ratios in those countries will naturally move in the right direction.

The Canadian statistics are intriguing for other reasons too, especially when the life expectancy issues discussed in the previous chapter come into play. Not only are more educated people entering entrepreneurship—about a third had university degrees—they're also getting older. People aged fifty and up were the fastest growing group of entrepreneurs, accounting for nearly a third of all new startups, with the affordability and availability of technology combined with well-developed skills and connections cited as the reasons. (Vampires!)

Startups—whether they're in Israel, Silicon Valley, Canada, or elsewhere—aren't like traditional businesses because they're not easily definable. Anyone who's had a great idea knows that it doesn't always come with an easy way to make money. Many entrepreneurs have to bend and flex those ideas, with numerous business models thought up and discarded along the way. Colin Angus, chief executive of iRobot—the company that makes the Roomba robot vacuum cleaner and the Scooba that unfortunately doesn't clean behind my toilet—once told me that he went through dozens of business plans. His original was to sell toy dinosaurs. But iRobot hit the jackpot by building bomb-disposal robots for the military before moving back into the consumer market with vacuums.[18] The principals of InteraXon, a tiny startup in Toronto that makes the Muse brainwave-reading headband mentioned in the previous chapter, originally thought their technology could control computers and televisions. They've since re-oriented to making meditation tools that connect to tablets. Every startup goes through this sort of trial-and-error process, often called a "pivot."

This necessary flexibility means it will be harder to pin down exactly what the companies of the future do and therefore what their workers do. It's already happening. Take Google, for example. Sure, it runs the world's most popular search engine, but it's also into maps, e-mail, operating systems, online storage, broad-band networks, robot cars, wearable computers, and life extension. What kind of company is Google again?

Big operations are having to work more like small ones in order to be flexible, with internal functions subdivided into self-contained units that essentially function as their own operations. Google is a particularly smart big company because it has its toes in a hundred different pools, thereby operating like a hundred different startups. It may make most of its money from search today, but who knows what the future will bring?

It's even more evident the further down the chain we go. Small businesses are having to morph and change, collaborating with one another and then going their own ways on an increasingly frequent basis. "It will be even more difficult to identify exactly what a small business is," says Tal. "It is also likely that domestic home-based firms will increase their partnership with non home-based firms to collaborate on individual projects. More small businesses . . . will become virtual corporations, stopping and starting on a project-by-project basis."[19]

On an individual basis, we've all heard how people these days have to have multiple careers. You may start as an accountant, but you probably will have to morph into an electrician or a computer programmer someday. That forced evolution and flexibility are accelerating to the point where people may have to change careers and "pivot" every day. As Brynjolfsson and McAfee put it, "the coming century will give birth to thousands of small multinationals with low fixed costs and a small number of employees each."[20] In other words, we're all becoming our own multifaceted corporations, or millions of mini-Googles.

## THE EARTHSHIP ENTERPRISE

Stronger economics and longer lives clearly are providing more people with the comfort levels they need to take the plunge and to try doing things that are personally satisfying. That says something profound about human nature. It shows that, when our basic needs are taken care of, we are naturally enterprising. This isn't unlike any point in history, but what's different now is that both push and pull effects are happening. Both opportunity and need—the encroachment of robots—are nudging more of us toward this true entrepreneurial nature.

This trend is having big social effects. A 2013 report from the United Way organization found that nearly half the people in and around Toronto, my home city, were "precariously employed," or working nonstandard jobs that are insecure and without benefits.[21] Union representation, meanwhile, is also down, as it is in the United States, where membership is the lowest it's been in nearly a century.[22] Entrepreneurship—by its very nature a form of "precarious" employment—explains some of this, but so too does the rising inequality discussed in chapter two. Either way, government policy will have to adjust to this push and pull. Modern economies are shifting away from the cozy setup of the previous paradigm, in which big companies employed people in secure jobs and supplied medical benefits while unions made sure workers' rights held firm. If those long-held standards erode, governments will need to adjust rules and regulations on a variety of matters, from unemployment insurance and availability of benefits for self-employed people to mortgage loans and pension contributions.

This all happened before, most notably during the Industrial Revolution. Inequality spiked then too, and entrepreneurs reaped all the riches while rank-and-file workers suffered. In response, unions formed at the grass-roots level and governments invested in education and therefore the preservation of the equality of opportunity. Similar mechanisms will need to manage the transition to a world in which robots and entrepreneurialism are the norm. Further investment in education seems like a no-brainer, but unions will have to evolve as well. Rather than representing workers in certain industries in certain countries, they may need to go global. As the epigraph to this chapter suggests, what has happened before is likely to recur. In the 1920s, unions largely organized themselves around specific trades, such as carpentry or glass blowing, which gave them power since they controlled guilds and apprenticeships. But their influence ebbed as big corporations arose and hired workers directly, which forced a reinvention. In the late 1930s, unions transformed and began organizing workers on a company level, rather than by skill category. They came to the key understanding that companies had to compete against one another. Any deals the unions struck with managements couldn't put those enterprises at a disadvantage. This system worked well in a non-globalized world.

Now, large corporations are competing in a worldwide market. Western companies are locked in battle with Eastern ones, many of which don't have the same cost structures, which is a natural advantage for those relative newcomers. As a result, the majority of union workers in Western countries today are employed by their respective governments, which are effectively the last bastions not open to foreign competition. Workers elsewhere, meanwhile, haven't seen wages or conditions improve much.

Today's unions must come to understand the global forces at work and then adapt to them. Just as businesses have gone global, so too must unions, thereby repeating the same sort of transformation they underwent in the mid-twentieth century.

The situation seems gloomy now, given the rising inequality and the stagnation of the middle class. If you read the daily news, it's easy to think pessimistically that the fixes to these problems will never come. But humanity has overcome issues like these before, and there's no reason to believe we won't do so again. We just might not be able to see the solutions yet. The technological tools enabling entrepreneurship are getting better and easier all the time; it's the public policies that inevitably need to follow. Human enterprise will continue to spread unevenly since not everyone will get these policies right. Some countries, such as Israel, will gallop faster into accepting this particular aspect of human nature, while others will inevitably take the slow road.

One of our greatest traits is our knack for enterprise, for thinking up new and better things or ways of doing things. That capacity has proved limitless so far, and it's not going to trail off any time soon. The suggestion that some people just aren't entrepreneurs—that only a small percentage of the population qualifies as such—is also probably not true. As we'll see in the next chapter, everyone is enterprising in some way and therefore has the potential to become an entrepreneur. If they aren't now, it's probably because it's not yet easy enough. If the average person won the lottery today, she might continue working at her dreary job just to fill the hours, but it's also a safe bet that she'd do something enterprising with the money; maybe a trip around the world or renovations to the house. But the ease and freedom from a windfall of money like that would inevitably spark her imagination in some way. Some people aren't as imaginative as others, but *everyone* is enterprising. They just need the means to express it.

# 5

# Arts: Long Live the Dead Buffalo

*Creativity is just connecting things. When you ask creative people how they did something, they feel a little guilty because they didn't really do it, they just saw something. It seemed obvious to them after a while. That's because they were able to connect experiences they've had and synthesize new things.*

—STEVE JOBS

The 2008 animated movie *WALL-E*, about the little trash compactor robot that discovers love, certainly deserved its Best Picture nomination at the Oscars. It offered proof that man and machine are capable of amazing creations when they work together. With its jaw-dropping computer-generated graphics, *WALL-E* was the perfect blend of technology and art.

But while it tells a positive story about how life and love can flourish in the darkest and most unexpected places, it's also quite cynical. In the movie, humans have abandoned life on Earth in favor of a giant spacecraft not unlike the mammoth pleasure cruise ships that currently prowl the oceans. They've left the planet behind because their pollution has rendered it incapable of sustaining life. WALL-E, the eponymous robot, is one of many machines tasked with cleaning up the mess while the people spend their days watching movies and playing video games in comfy hover chairs aboard their space cruiser. Prosperity and automation have reached their ultimate conclusions, so no one needs to do a lick of work, which is why everyone is pleasantly plump, their limbs stumpy and weak from disuse. In one of the funniest scenes in the movie, a roly-poly human gets up from his chair and tries to walk, only to stumble about like a youngster taking his first steps.

Funny, yes, but optimistic about our future? Not at all. The carefree, blob-like people offer us a cautionary tale, a warning that if we continue down our current path, we'll destroy the environment and end up as invalids, incapable of even taking care of ourselves. If that's not a cynical outlook on our future, I don't know what is.

But aside from its environmental and social commentaries, what I've always thought *WALL-E* got most wrong was its portrayal of our cultural consumption. Its humans have become invalids through their incessant watching of TV and movies and playing video games, which is an easy and even tired criticism to make. Who hasn't heard the complaint about how kids aren't reading or playing outside anymore, that they're spending all their time glued to screens? On top of that, the movie also suggests that people will increasingly become zombies, presumably under the control of whoever is piping out all those films and games that they're slavishly consuming.

Both suggestions run counter to what's actually happening. In the future, big, evil, faceless corporations or even governments won't be shoveling all that empty, distracting content into the trough of human consumption. It'll be us, ourselves. We're not just becoming hoggish consumers of content. Just as with our self-creating jobs, we're also becoming prodigious producers of it too.

## BEER NOT PHOTOS

In 1999, grunge music was all but dead, my hair had long been short, and everyone feared that the Y2K bug would destroy the world and everything in it. The digital revolution was just beginning; the Internet was starting to take off in a meaningful and life-changing way, and people were slowly discovering that all these newfangled electronics were much better than their old analog forebears. "Hey, look honey, the picture on this DVD really is much sharper than the VCR!"

For my first backpacking trek across Europe, aside from a guidebook and extra underwear, I'd also packed my Nikon SLR camera. Being in my mid-twenties and not made of money, I'd budgeted seven rolls of film, thirty-six exposures each, figuring that would be enough to get me through a month-long visit to some of the most amazing sights of the Old World. (For younger readers: film was a chemically treated plastic that we put in little canisters. Light etched the chemicals into images that we transferred onto treated paper, which ultimately gave us photo prints. Yes, it sounds like witchcraft to me too, now.)

So: 252 photos to cover Paris, Berlin, Rome, and Switzerland. No problem, right? After being extraordinarily stingy with what I actually photographed, the development cost me just about that number of dollars. But that wasn't even the worst part. With photography skills that bordered on nonexistent, perhaps a dozen decent images resulted from my efforts. I got a good Arc de Triomphe, a pretty decent St. Peter's Basilica, and a bunch of blurry Swiss Alps. The cost of the photos I kept averaged to about twenty dollars each. That seems insane now.

French inventor Nicéphore Niépce created the first analog photographs in the 1820s. The first consumer camera, the Kodak Brownie, wasn't released till nearly a century later, in 1900. The best estimates figure that inventors, tinkerers, and professionals took a few million photos in those intervening eighty years. Cameras got cheaper and better and finally went mainstream around 1960. About half of all photos taken that year were of babies, a good proxy for judging whether such a technology had reached critical mass with consumers.[1]

As we saw in chapter one, the first electronic camera—where the image was created and stored in digital format rather than replicated through chemicals on film—was invented at Kodak in 1975, but again, Moore's Law had to kick in. Kodak's device weighed nine hefty pounds and took twenty-three seconds to record a single black-and-white photo onto a cassette tape, which was then played back on a television screen with the aid of an additional console. Obviously, it was nowhere near ready.

Sony made a big breakthrough in 1981 when it developed a two-inch-by-two-inch floppy disc that replaced the tape, and digital cameras continued to improve throughout the decade. By the nineties, film's photo quality was still far superior and electronic cameras still cost around twenty thousand dollars. But the ground really started moving in 1990, when Switzerland's Logitech released the first true digital camera, a device that converted images to the ones and zeros of binary code, connected to a computer, and stored photos on a memory card.

That link to computers finally gave photographers an easy way to transfer pictures to their machines, where they could transmit them to others over the Internet. Meanwhile, Moore's Law took hold of other relevant components, including memory cards and image processors, so cameras quickly got smaller, better, and cheaper. Japan's Nikon kicked off digital photography for professionals with the D1 SLR in 1999. People had already taken a good number of photos—around eighty-five billion in 2000, or about twenty-five hundred every second. In 2002, 27.5 million digital cameras were sold, or about 30 percent of all cameras. The digital revolution in photography took hold fully in the mid-2000s, and the film market all but died by 2007, when 122 million digital cameras sold.[2] By the end of 2011, an estimated 2.5 billion people had digital cameras, whether SLRs, compact, or in their phones.[3]

With billions of cameras in existence, people took an estimated 375 billion photos in 2011 alone. As analyst and blogger Jonathan Good puts it, "Every two minutes today we snap as many photos as the whole of humanity took in the 1800s. In fact, ten percent of all the photos we have were taken in the past twelve months."[4]

The growth is perfectly logical. With the costs of film and development suddenly gone—no more twenty-dollar photos—people started shooting like crazy. Rather than carefully picking a shot, waiting for the perfect light, and framing it just right, they snapped away without a second thought. We used to get upset—sometimes violently so—if someone walked through one of those meticulously crafted shots just as we pushed the button, but now it was no longer a big deal. (To think: I could have spent that $250 in Europe on beer instead—okay, more beer—and still come home with more usable pictures than I did.)

Yet, this paradigm shift had a downside in its first few years. Many of the photos being taken were going straight into limbo. After all, what were we supposed to do with them? We could look at them on a computer, but who wanted to do that? It's one thing for a family to gather on the living room couch to pore over a physical photo album, but it was another to cram around a computer screen. It just wasn't the same. Intermediate solutions came along. Digital photo frames, for example, proudly displayed all those images we took, but by this point, a sort of *WALL-E* laziness had prevailed. Once, we had been perfectly content to finish a film roll, haul it down to the drugstore, wait a week for it to be developed, and then go back to pick it up. But somehow the act of transferring photos to a digital frame proved too wearying. As a result, the majority of those billions of photos taken never left the memory cards on which they resided; the memories they represented were doomed to never see the light of day.

But then the second part of the revolution kicked in. Along came Facebook, the social network that opened its doors to the general public in 2006. The website offered the perfect repository for all those photos because it actively connected friends, family members, classmates, and co-workers. These were the people with whom we actually wanted to share our photos, and Facebook importantly made the connections passive from the user perspective. You didn't have to search out a friend's photo album to see if you were in it, as had been the case on Friendster and MySpace. Your friend tagged you, you found out more or less instantaneously, and shortly thereafter you checked out the photos in question. Digital photography and Facebook thus evolved into a sort of symbiotic relationship. The website grew dramatically because users could see their friends' photos, while the number of photos on Facebook exploded because of the rapidly growing number of people on it.

The stats are almost overwhelming. In 2011, users shared some seventy billion photos on Facebook—about 20 percent of the total pictures taken in the world that year—bringing its total repository to 140 billion, or ten thousand times bigger than the photo catalog at the Library of Congress.[5] The number keeps growing ridiculously: In 2012, the website announced it was receiving three hundred million photo uploads every day, amounting to 109 billion for the year.[6] Plus, that's just Facebook. It's the largest driver of online photo sharing, but it's also just one of many photo-focused social networks. Flickr, Pinterest, Twitter, Facebook's Instagram, and Google's Picasa all contribute their share as well. As more and more people come online every day around the world, the exponential growth is only going to continue.

The photo explosion helps us understand how technology is changing and shaping human culture itself. For much of the past hundred years, photography was a communication medium, hobby, and profession reserved for those who had the money to engage in it—twenty-dollar photos!—as well as the patience to learn its techniques properly. Technological advances eventually removed the cost and skill needed to take decent photos, then made it easy to distribute and share those images with their intended audiences. Now, with the barriers gone, photography is no longer the domain of a small elite group of people. It's a communication medium on a grand global scale.

Sure, we can talk, e-mail, and text, but like the cliché says: A picture is worth a thousand words. An image can convey much more than simple text can, which is why people share photos of sunsets, concerts, and the artisanal ham sandwiches they had for lunch. Some people dump every photo they take onto social media, though most deliberate on their choices a little more. Each image we share, after all, does say something about us. Even the humble ham sandwich can tell viewers about its sharer: that he doesn't keep Kosher, that she has a fetish for gourmet comfort food, or that he's lonely and wants someone to pay attention to him. Whether intended to amuse, entertain, challenge, or inspire, most communicated photos are a conscious form of human expression. It may not be art or even good, but every shared photo conveys a series of messages.

## THE MESSAGE IN THE MEDIUM

Photography is the most obvious medium to benefit from technological advance, but it's hardly alone. Virtually every means of human expression and communication has grown significantly, especially in the past few decades, because of ongoing advances in production and distribution.

In the Middle Ages, literacy was reserved for the rich and privileged. Prior to the 1450s, when a German goldsmith named Johannes Gutenberg came along with his printing press, a small handful of monks safely tucked away in their monasteries jealously guarded the sum of human knowledge from the filthy masses. Gutenberg's invention paved the way for mass literacy, and, over the next few centuries, more and more people became readers. Published books expanded from religious texts to great works of literature, then eventually to the likes of Dan Brown, Stieg Larsson, and *50 Shades of Grey*. By 1990—at which point about three-quarters of the world's people could read—publishers were pumping out nearly a million books a year to meet demand.[7]

While total world population has grown by a third since the early nineties, the number of books published has comparatively exploded by more than 150 percent in the same time frame. Continued growth in literacy—the global rate climbing to 84 percent by 2010—fueled some of that increase, but globalization has played a major role as well. The United States, United

Kingdom, France, and Spain have become especially important because they publish in languages used internationally in former colonies, with most major multinational publishers consequently based in those countries. The United Nations Educational, Scientific and Cultural Organization (UNESCO) says those four countries "constitute the main international centers of publishing and have considerable influence beyond their borders."[8] Naturally, the numbers of publishers in those countries have proliferated. In the seventies, about three thousand publishers were operating in the United States, and by the end of the twentieth century that number had increased to sixty thousand. Pessimists claim blindly that no one reads anymore, but the number of titles published has grown dramatically worldwide, to about three million in 2010, a tripling in just twenty years.[9]

The revolution in self-publishing, meanwhile, is only just beginning. As with the rise of digital photos, two entwining arms of technology have made it easy for anyone to write a book. The first arm is the personal computer, a capability we've had at our disposal for the better part of thirty years. The second arm is micro-distribution in the form of print-on-demand technology and controlled electronic publishing. In the latter instance, the necessary advances have emerged only recently. Online retailer Amazon has been the biggest player in that area so far, having launched its Kindle Direct Publishing platform in 2007. Not only does the system allow authors to sell their work directly to consumers without having to go through a publisher, it typically lets them keep most of the proceeds as well. With the simultaneous proliferation of e-readers and tablets, a new world has opened both for established and would-be writers, many of whom jumped at the chance to reap more of their own rewards. Not surprisingly, self-published books—once the exclusive domain of losers who couldn't get a "real" publishing deal—are skyrocketing. In the United States alone, self-published books grew nearly 300 percent between 2006 and 2011 to a quarter-million titles. In the summer of 2012, four self-published authors had seven novels on the *New York Times* bestseller list. As author Polly Courtney, who went back to self-publishing after three novels with HarperCollins, put it: "It feels as though the ground is shifting at the moment . . . It's quite liberating. Some sort of transition was overdue."[10]

But books aren't the only medium for writing. The Internet has made it possible for anyone to start an online diary, commentary pulpit, or even news organization. The number of blogs has mushroomed from thirty-six million in 2006 to more than 180 million in 2011. The activity is proving particularly popular among women and young people, and a lot of people are clearly reading all these blogs. More than a quarter of all Americans online regularly check out the top three blogging websites: Blogger, WordPress, and Tumblr.[11] More people are writing, and more people are reading. Nearly a billion people in the world still have yet to crack open a book or see a blog, but literacy rates are improving.[12] Reading is far from a dying pastime; its future looks bright.

## INDIE GOES MAINSTREAM

It's a little trickier to track growth in the music industry. Until relatively recently, no one could record performances of music, so the history of the music industry was the history of music publishing, from the huge tomes of religious music copied in the Middle Ages to the sheet music printed and published into the nineteenth century.

Recording sound first became possible in 1857 when Édouard-Léon Scott de Martinville came up with what he called the phonautograph. The machine tracked sound waves moving through the air and replicated them using a stylus on a sheet of soot-coated paper. It couldn't play the sounds back, but the process was soon reversed into a new device that could. Thomas Edison perfected the phonograph or "record player" in 1878, and so was born the era of popular music distribution. By the turn of the twentieth century, three companies—Edison, Victor, and Columbia—were selling about three million records a year in America alone.[13]

Decades before Moore observed the phenomenon, the technology advanced, improved, and got cheaper, to the point where an industry formed around it. Music became a business, with recordings moving from wax cylinders and discs to vinyl albums, cassette tapes, compact discs, and ultimately to digital files.

During the first half or so of the twentieth century, hundreds of record labels sprang up around the world to help produce and distribute music recordings for profit. In the latter half of the century, many of those independent operations consolidated into a few mega-players. By 1999, five major corporate entities essentially controlled the industry: Germany's Bertelsmann AG, Britain's EMI Group, Canada's Seagram/Universal, Japan's Sony, and US-based Time Warner. The concentration of power had its benefits. It lowered infrastructure and transaction costs, gave musicians greater access to the sorts of skilled and specialized workers they needed—studio technicians, producers, marketers, and the like—and introduced the ability to monitor and learn from the competition.[14] The major labels controlled every aspect of music production and distribution, from the recruitment and development of artists, to legal services, publishing, sound engineering, management, and promotion. On the downside, it became incredibly hard and rare for an individual to produce and distribute music without having to go through one of the big gatekeepers.

As such, the big companies enjoyed tremendous success. In the United States alone, music sales soared 160 percent between 1987 and 1997 to $12.2 billion, surpassing every other medium. As researcher Brian Hracs puts it, "In 1997, the recorded music sector stood on top of the entertainment pyramid, surpassing domestic sales in the motion picture industry, as well as DVDs, video games, and the Internet."[15]

Then, the so-called MP3 crisis hit, sending the industry into a tailspin. The new format allowed for the digitalization of music into small files, easily transmitted over the Internet. A file-sharing gold rush ensued. Napster in particular capitalized on the phenomenon. In 2000, half a million people were logged into the file-sharing service at a given time, and a year later that number had ballooned to sixty million. With people copying and acquiring music for free, industry revenues plummeted. Global music sales fell by 5 percent in 2001, then by a further 9 percent in 2002. One estimate pegged the decline in consumer spending in Canada between 1998 and 2004 at a whopping 40 percent.[16] For the music industry, the sky was falling.

Yet, despite nearly a decade of MP3s, file-sharing, and iPods, music production continued to rise. Between 2000 and 2008, the number of new albums released in the United States more than doubled to 106,000, according to tracking firm Nielsen.[17] Releases took a dip after that—to seventy-five thousand in 2010, partly explained by the global recession—as record labels finally adjusted to the new musical reality. It took a while for the full effects to take root, but the MP3 decade forced labels to become considerably more risk-averse. Now, they release albums mainly from proven money-drawing acts. Labels have transformed from their lucrative days in music development and production into brand-led marketing companies akin to Coca-Cola or Nike. Artist development, which they used to oversee, now falls to others, primarily the artists themselves.

With the big boys taking fewer risks, the business has now become much more independently driven. In the United States, the number of label-employed musicians declined by more than three-quarters from 2003 to 2012.[18] In Canada the number of musicians estimated to be without a recording contract and therefore technically indie is a whopping 95 percent.[19] Those figures suggest that the traditional means of tracking music production and distribution—primarily Nielsen's album releases and sales—no longer have much relevance, at least not as far as gauging the real state of the art.

Some estimates peg the number of indie albums released in the United States in the realm of ninety thousand, or greater than the number released by actual labels.[20] Even counting albums has become something of an outdated method since single songs have become much more popular in the digital age.

It's difficult to track and estimate exactly how much music is being created and distributed, but with musicians having access to continually improving tools to make and disseminate recordings—whether it's editing software such as Pro Tools or online stores such as iTunes—the medium has likely experienced the same big production growth as photography and writing. "With computers and file sharing, a whole lot more people are making [music]," says Anthony Seeger, professor of ethnomusicology at the University of California Los Angeles. "They can post their own latest composition, their own latest song. It's stimulus to compose and write and get it out there and it's also a mechanism that allows more people than ever possible in the past to do so."[21]

## OPTIMUS IPHONE

The same is happening in the movie industry. Here too the dual forces of globalization and technological advance have had a major effect both on production and distribution. With a potential worldwide audience, studios have been increasingly cranking out feature films, nearly tripling their production between 1996 and 2009, from 2,714 a year to 7,233. That growth also covers both developed and developing countries.[22] In the United States, for example, 429 films were produced in 1950 versus 734 films in 2009. The comparative numbers in India were 241 versus 1,288.[23] Many people would consider Hollywood the world capital of moviemaking, but it's not. Both India and Nigeria surpass the United States in total films produced, with Japan and China rounding out the top five.

Just as with the music industry, technology is driving a major shift in how movies are made and distributed. With bigger and better televisions and increasingly affordable home theater systems, Hollywood studios are undertaking larger-scale productions that need to be seen on giant screens to be enjoyed fully.

Big blockbusters therefore are getting more expensive to make. The 1994 Arnold Schwarzenegger action flick *True Lies* was the first movie to cost $100 million, but that seems cheap considering that thirty films between then and 2013 cost double that or more. Conversely, cheaper cameras, sound and lighting gear, and editing tools are making it easier to make movies on the low end. If a budding director can't afford the already cheap professional quality Red Camera, which cost only twenty-five thousand dollars in 2013, he or she could always enter the iPhone Film Festival, where the only major outlay is generally a two-year cellphone contract.

More important than the hardware advancements is the proliferation of alternative distribution options. Cable TV, YouTube, Netflix, and a host of other online services are giving filmmakers more and cheaper ways to showcase their work. "Cinema is escaping being controlled by the financier, and that's a wonderful thing," says Academy Award–winning director Francis Ford Coppola. "You don't have to go hat-in-hand to some film distributor and say, 'Please will you let me make a movie?'"[24]

Not surprisingly, indie production—just as in music—is booming. The Sundance Film Festival saw submissions increase by 75 percent to more than four thousand between 2005 and 2011, while over the same time frame Cannes saw its submissions balloon by 184 percent to 4,376 a year.[25] Filmmakers whose pictures major distributors aren't picking up are finding willing buyers in cable companies' video-on-demand services, Netflix, and a host of other online streaming operators.

But, as with writing, big industry isn't the only outlet for budding filmmakers. YouTube has emerged as a platform on which creators can display their work and even create a profitable business. The growth there has been even more astronomical. In 2012, the site saw one hour of video uploaded by users every second, up from "only" ten hours every minute in 2008. It registers four billion views a day and three billion hours of video watched every month. More video was uploaded to YouTube in one month than the three major US networks created in sixty years.[26] The reason for this growth is obvious: People like to communicate with each other, and video has become one of the more popular mediums on the Internet for doing so.

YouTube has also become an important launching pad for undiscovered talent. Patrick Boivin's route to success is remarkable. As a teenager in the 1990s, he was pretty typical. He didn't like school, but he did love drawing comics. Whenever he thought about his future prospects, though, he resigned himself to the likelihood of working as a waiter or some similarly low-paid drone job to support his passion. But he was enterprising, so he got together with a few like-minded friends and made some sketch-like short videos that spotlighted their artworks on VHS tapes. They showed the videos in bars on weekends and they proved to be popular. The tapes drew the attention of a small local television network, which hired Boivin and his crew to create a series called *Phylactère Cola*, a sketch show that parodied and otherwise skewered movies, television, and society in general. The show aired in Quebec in 2002 and 2003 and served as a sort of school for Boivin, who used it to learn the ropes of filmmaking. When it wrapped, he bounced around jobs, making commercials for various clients.[27]

Around this time, YouTube began its meteoric rise. Eager to draw the millions of viewers that some uploaders were getting, Boivin posted some of his short films on the site. But the big audiences didn't come. He wasn't completely disheartened, though. He studied what made popular videos go viral and set off to do the same. Some people were doing well by reconstructing scenes from the *Transformers* movie with the toys. The movie was hot, and people were talking about it. He figured that he could take advantage, given his background. He created a stop-motion video of an animated fight between two of the robots, and when that went viral he made more. Further success got him thinking about how he could make a living from this sort of work, and he came up with the idea of selling his services to toy companies. "Eventually, it worked," he says. He's been making viral videos as a full-time job ever since, with YouTube serving as a sort of audition tape for employers.

"It became a different way of making a living as an artist, which is amazingly rare because you usually don't have many options, especially as a filmmaker," he says. More importantly, there's the opportunity of exposure. Before such widely available platforms existed, filmmakers often didn't act on the ideas they had because they weren't sure if they would ultimately be worth it. "It became an incredible motivation because the hardest thing I experienced as a movie maker—all the movie makers I know experienced the same thing before YouTube—to find an audience was so hard that most of the time you didn't really care about doing something because you thought it would only be seen by a couple hundred people," he says. "When you know that doing something, when it's good, will be seen by thousands or millions of people, now that's something. The days we are living in now are special because you know that what you do can be seen by the whole world."

In 2012, Boivin uploaded his first feature film, the sixty-eight-minute *Enfin l'Automne* (*Fall, Finally*), just to see if he could do it. When we spoke, he said he was on the verge of signing a deal with a Hollywood studio to do a proper film. In 2013, he was tapped to direct *Two Guys Who Sold the World*, a sci-fi comedy that won a lucrative film pitch prize at the Berlin International Film Festival. Like the millions of people of varying levels of creativity who have uploaded and shared photos, e-books, blog posts, songs, and videos, Boivin owes his livelihood and success to the technology that made it possible. He speaks like a pitch-man for YouTube, but he's actually selling the virtues of technological advancement. "As a French Canadian, it's an opportunity that didn't exist before. And it's all thanks to YouTube."

The same has held true for me, albeit in textual rather than visual context. I started a blog in 2009 to promote my first book *Sex, Bombs, and Burgers*. Over the years, it morphed into a broader canvas where I shared my thoughts on technological trends, news, and events. After a few years, in which I gained something of a following, I signed a deal with a pair of magazines to syndicate it, which provided me with a steady source of income. More importantly, the blog has served as my online portfolio and chief marketing tool and has landed me numerous jobs and opportunities. It's an incredibly important platform that makes possible my own efforts at entrepreneurialism, as that kind of thing does for many others who write for a living.

## AN EXPLOSION OF PLAY

Often derided by critics as passive or even mindless, anybody who actually plays video games knows that they're anything but a medium devoid of cultural value. Games are at least as valuable an art form as movies, and they engage players in ways that no other medium does. In games with narratives, for example, players make choices that affect the outcome, which means they become more invested in the characters and stories they create than if they were only passively watching. Games often take the brunt of the blame for lulling children into sitting in front of a screen for hours—perhaps to become fat blobs as in *WALL-E*—but they're no worse at doing that than other similar activities, like watching movies. Video games have become the scapegoat for every social ill, from overweight children to school shootings, largely because of their relative newness as a medium. The first commercial video games became available only in the early 1970s, making the medium a child relative to other forms of entertainment, all of which went through similar growing pains during their own respective infancies. Radio and television were similarly demonized for spurring social ills in their early days.

The negative attitudes are starting to ebb as better and cheaper technology is expanding the appeal of the medium beyond its initial audience of younger people. Games used to be played primarily by kids on computers and consoles in dark basements, but in recent years they've proliferated onto smartphones and tablets. Many adults, some of whom grew up with Nintendo or Atari consoles, now while away their commutes by playing bite-sized games on their phones and tablets. Video games are no longer just ultra-violent, forty-hour shooters with incredible graphics that cost sixty dollars; they're also quick distractions that can be had for ninety-nine cents or less. The medium is maturing, and gamers are getting older on the whole. The average age is now about thirty.[28] The raw number of people playing worldwide has risen from 250 million in 2008 to 1.5 billion in 2011.[29]

In the early 1980s, the predominant way to play video games was on the Atari 2600 console, which had a total catalog of just over four hundred titles.[30] Since then, platforms and distribution methods have proliferated. Now, designers can get their games to a multitude of devices via an Internet connection, whether it's home broadband or cellular. They don't necessarily have to go through a gatekeeper like they used to, which in the past meant console makers such as Atari, Nintendo, Sony, or Microsoft. By way of comparison, the number of games available for Apple's iPhone alone hit more than six thousand in 2009, less than a year after the company launched its app store, and climbed to ninety thousand by the end of 2011.[31] The growth has been similar to that of photos: More games are being released for just the iPhone every three days than were created for the entire run of the Atari 2600.

The number of game makers is correspondingly exploding. In the United States alone, eighteen thousand game-related companies were operating in 2008, or eighteen times the number that existed just three years prior.[32] Just as in film, where the budgets on the biggest games continue to go upward, it's now equally possible for smash hits to be created on a veritable shoestring. *Minecraft*, developed and released in 2011 by a single person—Swedish programmer Markus Persson—sold more than a million copies before its full version was even released, netting its designer more than thirty million dollars. It was only after the game saw major success on its own that Microsoft asked Persson to make it available on the Xbox 360 console.

Even with its staggering growth, video game production still remains very much the domain of a relatively small elite group: the highly skilled computer programmers who know how to make them work. But that is already changing. The next step in this proliferation of production is for game creation itself to go mainstream. A growing number of titles are making so-called "user-generated content" a key feature by which players themselves dictate large portions of the game. It started with the ability to customize the looks of in-game characters—a different-colored hat here, a different shade of skin there—but has evolved into creating entirely new characters, levels, and worlds, as in *Minecraft*. The idea is for players to share their creations the same way they would share photos or music files, which can greatly increase the re-playability of a game and therefore its value. It's happening because, once again, the tools to do so are becoming cheaper, better, and easier. Games are increasingly allowing their players to become designers without having to know the first thing about computer programming.

## BIG LITTLE PLANET

One of the best examples is *LittleBigPlanet*, a game released for Sony's PlayStation 3 in 2008. Starring a little puppet known as Sackboy, the game was cut from the same cloth as rival Nintendo's *Super Mario* titles, which typically require players to jump from one platform to the next while gathering various treasures and power-up items along the way. *LittleBigPlanet*'s twist on the traditional formula came from giving players the ability to design and share their own levels online. With a straightforward, easy-to-use toolset, *LittleBigPlanet* became a phenomenon. Not only did the game sell more than four million copies in its first two years and win all sorts of critical accolades, it also saw more than seven million levels published online by 2012.[33] Players who bought the game theoretically could play it forever. The only limit to the number of levels was their own imaginations.

You'd think the game's maker, Media Molecule, would be something of a legend in Guildford, England, the small town about thirty miles southwest of London that it calls home, but you'd be amazingly wrong. The building by the train station does house some multimedia companies, but nobody there has ever heard of Media Molecule. The receptionist does some quick Google searching and sends me off to the center of town. Her address, however, turns out to be a parking lot. An older gentleman notices my confusion and offers to help, but he hasn't heard of the company either. Nor has a nearby police officer, who surely must know of all the important local businesses, right? Nope, he has no idea who or what a Media Molecule is. He points me to the nearby offices of Electronic Arts, the biggest video game maker in the world, and I breathe a sigh of relief. A fellow game company undoubtedly will know where its competition is . . . which is why it's so astonishing that no one there does. The receptionist has never heard of the rival developer either, nor have a handful of EA employees who overhear our conversation. By this point, I'm convinced I've journeyed to the wrong town.

Finally, one helpful fellow—probably noticing my growing exasperation—comes over to the reception desk and reveals that, yes, he has heard of the *LittleBigPlanet* maker. He goes upstairs to his desk and retrieves a business card, which has a phone number. We call and—eureka!—finally discover the long-lost land of Media Molecule. With the correct directions in hand, I walk the three blocks to the studio and let fly a stream of curses upon discovering that it *is* right by the train station.

Despite creating some highly successful games and being acquired by Sony in 2010, Media Molecule has purposefully stayed small. The comedy of errors in finding the place is ultimately worth it, with the studio turning out to be uniquely eccentric. The carpet running the length of the main floor is hot pink, which is enough to startle tired employees into a state of readiness as they trudge into work in the morning. The common areas—meeting rooms on the first floor, and kitchen and recreational room on the second—ooze relaxation with a funky, seventies decor. Green and pink throw pillows sit atop mismatched patterned blue and floral-print couches, and the beanbag wouldn't look out of place in a kindergarten classroom.

With only forty employees, it's a fraction of the size of giant operations like Ubisoft Montreal, for example, which employs thousands. The smallness, along with the low profile, explains why almost no one in town has heard of the place. Each employee therefore has lots to do; there are no single-task jobs here, like "background artist" or "debugger." Everyone wears a number of hats, which keeps them busy and ensures their workspaces lie buried in clutter. During my visit, the team is ramping up production on *Tearaway*, a new game project for the handheld PlayStation Vita system that lets players make their own character creations, then print them out as paper models. The office is buried under prototype creations.

David Smith, the studio's co-founder, technical director, and lead designer, greets me in the meeting room. He looks like the stereotypical video game maker: He's wearing the standard uniform of jeans, sneakers, and a gray T-shirt, with glasses and a pulled-back ponytail. He probably would have played Dungeons & Dragons with me when I was younger, a feeling he corroborates when he starts talking about how he loves both art and math. He tells me he's lucky that he gets to work in video games, the intersection of the two subjects. "I get to scratch both those itches: the geeky technology where new things are always happening, but also the new experiences that we can make and that make us laugh," he says. "I really do feel truly blessed to be part of this surging wave, this crazy evolution that's going on right now. If I had been born ten years earlier, I'd probably have gone down an engineering route and probably felt that the creative side of me was a little bit suppressed."[34]

He nods knowingly as I explain the premise of *Humans 3.0*. I tell him about the chapter I've just finished writing, about how entrepreneurialism is on the rise. He's not surprised. "It's another form of playfulness," he says. Business creativity coincides with the growth of art, communications, and expression, and all of these different aspects speak to people's desires to create, experiment, and share. All of them also use the combinatorial aspect of creativity to generate ideas continually and exponentially.

Humanity is at a watershed moment in its history, he says. We're emerging from a prolonged age of darkness, and he uses the analogy of playing games—to be expected, given his line of work. As children, we used to spend all day playing, constantly experimenting and making up new games. I immediately remember how my friends and I came up with our favorite game, a combinatorial variation of tag, tennis balls, and prefabricated playgrounds. Every child everywhere has thought up similar creations, yet at some point we all stop doing it because of the encroaching commitments and responsibilities of adolescence and then adulthood. As we age, we acquire bills to pay and dependents to support. "You can't afford to mess around," Smith says. "You have to be the bread winner and kill the buffalo."

Yet, if all the creation, sharing, entrepreneurialism, and gameplaying by adults are any indication, a profound shift is taking us back to that childhood experimentation. Our increased prosperity and longer lives are giving us the means and the time to take more chances, so we don't necessarily have to focus all of our energies on killing the metaphorical buffalo. The encroaching realities of the changing job market—in which we're all likely to change careers several times thanks to those damn robots—also means that we have to keep learning longer into life. Figuring out new skills and capabilities means experimenting. "You can't get a job at the bank and know that fifty years later, you'll die at the same place or be pensioned out," Smith says.

The financial matters that are often tied to discussions of this explosion in sharing become irrelevant in this larger context. Who gets paid and what individual copyrights are violated don't seem nearly as important as the large-scale shift toward rediscovering our ability to experiment and create. At the same time, the huge explosion in supply of art—music, writing, photos, and movies—seems to have devalued the demand for it. Many people are less willing to pay for the good stuff when free so-so stuff will do. But this is only a temporary, technological fluctuation that society will eventually level, Smith says. Having gone through the process of creating itself and therefore acquiring an understanding of what it requires, people may in fact become more willing to pay for really good art. "What you can't cheat is that it takes time to make something that is unique and good. If it took you only ten minutes to make something, I really think you can't make something that sounds truly new."

Smith at once feels a sense of accomplishment and of community in regards to *LittleBigPlanet*, knowing that what he made is now more important to other people than it is to him. The bean counters who sign the checks may not share that view, but it's a historical force that's in unstoppable motion. "Even though I'd spent a lot of my energy making this creative environment, it wasn't really something that I happily sat inside. There's all these people who love that space—it's now theirs. It's out of my hands," he says. "There's a certain switch-over point where we realized that the lunatics are running the asylum."

The future of art, entertainment, communication, and expression, meanwhile, is dazzlingly bright. The flood of content is ever increasing and therefore competing with itself, which may be harder for any one individual to get attention, but history suggests that this situation won't discourage people from trying. For the artists and the communicators themselves, the need to express themselves drives them in the first place.

"I wouldn't be bothered to make something cool unless I thought someone else would see it, even though I like to think

I create things for their own value," Smith says. "On some level, I kind of think this will help me communicate with someone or someone will see that and there will be some kind of human expression. If that was thrown away, I don't think I could do that. It's not something you consciously think about, it's just part of being human."

Which confirms my suspicion. The dual explosions of entrepreneurialism and personal expression indicate that technological advance is bringing people closer to their true natures as far as creativity is concerned. The need to create and collaborate flows in our blood. That doesn't apply to just a segment of the population but rather to all of us. If some individuals aren't doing that yet, it's only because it's not yet easy enough for them to do so. Like rice on a chessboard, it's only a matter of time.

# 6

# Relationships: Superficial Degrees of Kevin Bacon

*My wife and I have the secret to making a marriage last. Two times a week, we go to a nice restaurant, a little wine, good food . . . She goes Tuesdays, I go Fridays.*
— HENNY YOUNGMAN

All of this communication and expression are great, but what does it mean in terms of how we're actually getting along with one another? After all, what we say—whether through music, video, photos, or any other medium—is pointless if it doesn't make an impression on the intended audience. Are we just blowing a lot of smoke into the ether or is this technologically driven surge of communication having any effect on how we relate to one another? Are we getting along better or worse?

It's a big question because many types of human relations exist, all governed by an equally large number of political, economic, social, and cultural divisions. On the most macro of levels, there are international and intra-national relations—how different countries and the people within them get along with each other—which we discussed in chapter two. Moving downward, we have more micro-level relationships, how we interact with our families, friends, and lovers. These aspects, and how technological advances are affecting them bear investigating because collectively they comprise much of what it means to be human.

Let's start with the unit most of us interact with every day: our family. On a basic level, technology is having the expected effects and side effects on how family members interact. Smart-phones, for example, have made it easier for parents to stay connected with kids and know where they are at all times. Video calling has made it possible for individuals across generations and locations to stay in meaningful touch. In their marketing materials phone and tablet makers often show grandparents bridging vast distances with their grandchildren through the magic of video conferencing. On the downside, this proliferation of communications technology has led to an ironic disconnection: Picture a family sitting together in the same room, but rather than talking to one another everyone is staring instead at a backlit screen. We'll delve into this particular issue over the next two chapters.

But technological progress is changing the very structure of families on a much deeper level. They're transforming into something not seen before. As mentioned in chapter three, couples in the most advanced countries are having fewer children, meaning that families are becoming smaller. More than half of all households in developed countries today have no children, while a fifth have only one. Multi-children households are a shrinking minority: fewer than a quarter have two or more.[1] The Organization for Economic Cooperation and Development (OECD) reports that the child population across its member states is decreasing steadily, with the shrinking particularly pronounced in Germany, South Korea, Poland, and Slovakia between 2002 and 2010. For most of human history, families typically have been rather large, so this trend is unprecedented.

Moreover, youth dependency—the ratio of people under working age versus those above—has also been in steady decline across the OECD for the past sixty years. That's good because a lower ratio means people and the economy at large don't have to work as hard to support everyone. Mexico's ratio in 1970, for example, was 150 percent, meaning that each person had to work hard enough to support one-and-a-half people; the country's ratio is now only about 70 percent. Germany's ratio, in comparison, was 50 percent in 1970, but is now around 30 percent.[2]

Demographic changes coupled with economic improvement are also having big effects on family relations. China has seen a decades-long economic boom unfold alongside the effects of its one-child policy. The policy was intended initially as a population-control measure, but it has given researchers a wealth of data on how economic improvement correlates and interacts with smaller family sizes. China's experience in this regard can be seen as pre-scriptive or even predictive as to how the future is likely to unfold in developed countries, which are essentially arriving at their own one-child (or fewer) phenomenon through natural means.

China's one-child policy has become famous—some say notorious—for producing "Little Emperor Syndrome," the spoil-ing of children to the point where they become whiny, entitled brats. Parents are becoming increasingly well off and don't have a multitude of children to support. Many find themselves with a lot of disposable income. The children, who don't have siblings, want it all. An element of this development has trickled over in ironic fashion to the West, where social media users regularly joke about their own "first-world problems." A hilarious repository of these has sprung up on websites such as First-world-problems .com, where people share complaints like, "I have to get dressed so that I don't look too lazy when I go out to pay the gardener," and "I want to enjoy my beer in the garden but the Wi-Fi doesn't work out there." Plenty of us in the West already suffer from a kind of collective Little Emperor Syndrome.

In more than two hundred scientific studies in China, how-ever, no real differences—for good or for ill—have been found between only and non-only children. Single children tend to score better in sociability, competition, and cognitive self-estimation, but they don't fare as well in taking care of themselves.[3]

Researchers have also found that parents tend to spend more time with only children. In one study in Beijing, more than three-quarters of mothers and half of fathers spent at least half their leisure time with their one child, versus 60 percent and 45 percent, respectively, for parents with multiple children.[4] Evidently, the more children a couple has, the less time they want to spend with any of them. It's a qualitative assessment to say so, but most people would agree that the more time a family spends together, the better, so this is probably a positive development.

The Chinese demographic trends also point to a growing issue, however, on the opposite end of the age spectrum. In days of yore, an increasing number of old people wouldn't have posed much of a problem since multiple generations commonly lived together. But with fewer children being born and with people in general becoming long-lived vampires, there's a double pinch happening: There are fewer young people to take care of more old people. A 2008 study on China's one-child policy sums up the problem, endemic to all countries heading in this same direction: "A young married couple has to take care of four parents without help from siblings. If the one-child policy persists, the care burden of second-generation of only-child couples would be doubled or even tripled and the pool of family support of aged parents would shrink."[5]

That increasingly well-off young people today don't want to live with their parents and grandparents, like they once did, is compounding the issue. As the report says, "Increasing independence and increased availability of housing has also been a force in changed separate living arrangements. Young people often do not want to co-reside with their parents even when they live in the same city."[6] In China, that means attitudes are shifting toward how old people are cared for. Nearly a quarter of people between the ages of twenty and forty expect institutional care once they hit their golden years versus only 11 percent of those over sixty. Older people expect familial support, probably a vain hope in a world where Little Emperor Syndrome and first-world problems are on the rise.

It's not just a Chinese problem either. Attitudes in China are merely catching up to what Western countries have been experiencing for a while now. In the United Kingdom, for example, between a third and a half of seniors lived in a household with at least one of their children in the 1960s. By the 1990s, that figure had declined to between 5 and 15 percent.[7] In the West, we've long been shunting off our older generations into nursing homes and seniors' apartments. It's even become a punchline in the form of Grandpa Simpson on *The Simpsons*, who pleads with his family to let him into their home because "it's cold and the wolves are after me."

The separation is a shame because, despite current and future generations sharing every detail of their lives through Facebook, Twitter, YouTube, and the rest, oral history—a rich human tradition since the times of Blarg and a large part of who we are—may be falling through the cracks along the way. Technology and especially commerce tend to focus on the most popular or trendy information or stories, but whole chunks of regional and local cultures don't face the same commodification and thus aren't distributed as widely, if at all. In other words, we might be ignoring what's in front of our noses in favor of what's going on across the country as we stop having those close, familial conversations, where such information is discussed and passed on.

Surveys show that both old and young people strongly believe they can learn from each other, yet ironically they also say they have a hard time relating. In the same polls, all generations believe that social disengagement with young people comes naturally with aging. That attitude isn't surprising given the diffusion of media over the past few generations. A typical teen growing up in Britain in 1960, for example, had two television channels and a few radio stations for information and entertainment.[8] Today, the choices are nearly infinite, which makes for a lot of exposure to different cultural ideas. Put all of this together, and it's no wonder that young people laugh at old people who try to remain hip to the current jive.

The solutions to these generational disconnects may be quite simple. The sort of ubiquitous technology discussed at the beginning of this book could make things much easier . . . literally. With technology—or at least the best, most successful examples of it, such as Apple's iPad or Nintendo's Wii—becoming increasingly easier to use, older generations may find it a little less difficult to get "hip to the jive" and therefore reconnect with those crazy whippersnappers.

The overall trend is telling. As the Chinese experience illustrates, with family sizes shrinking and economic prospects improving, people are showing a stronger predilection for insularity. The amount of communication and self-expression is exploding, but we're voluntarily and perhaps even consciously shrinking our immediate familial circles. This tendency is even more true when it comes to those we consider our friends.

## ALONE IN A CROWD

If social media metrics are to be believed, we've never had more friends. In 2011, Facebook reported that the average user had 190 of them. A year later, the Pew Research Center found that number to be even higher, at least for Americans, at 245.[9] Facebook indicates that the six degrees of separation theory first put forward in 1929 by Hungarian author Frigyes Karinthy—where everyone on Earth, including Kevin Bacon, can be linked to everyone else in just six simple connections—is shrinking. In 2008, the average number of links was 5.28 and by 2011 it was just 4.74.[10] If we can take the website at face value, it's succeeding in its mission of bringing people closer together.

But we all know that's bull because nobody really has two hundred friends. You probably don't even know half the people labeled as your "friends" on Facebook. Maybe you met some of them once at a party a couple of years ago. You may have gone to elementary school with others. You've probably accepted friend requests from a few random strangers because their profiles looked interesting or attractive. Facebook calls them "friends," but that doesn't mean they are. You generally can't borrow a hundred bucks from most of your Facebook friends, nor would you trust them to feed your pets while you're away on a trip.

But these online services masquerading as social networks are very good at creating opportunities to make new friends, which shouldn't be undervalued. Never before have people been able to connect with like-minded individuals so easily. Do you have a maniacal interest in music that sounds like it was made in the 1980s but wasn't? You're not alone. There's a group for that. Fascinated by obscure BBS boards? Yup, there's a group for that too. The Internet is a wonderful place for people of disparate interests, backgrounds, and geographies to connect, and from there they can see real, actual friendships develop.

But can we really have hundreds of friends? In the 1990s, British anthropologist Robin Dunbar famously came up with "Dunbar's number," which posits that our brain's cognitive ability to remember everyone and why they matter to us limits the number of stable relationships we can have. Experts generally peg the number of relationships that we can maintain somewhere between 100 and 230, with 150 cited most often. Dunbar also believed that social networks resemble concentric circles. While 150 may constitute the outer boundary of friends, we subconsciously limit our group of trusted friends to about fifty, good friends to fifteen, and best friends to five. Some online social networks, including Path and FamilyLeaf cater to this idea by limiting connections to those smaller circles.

Sites such as Facebook, on the other hand, make it easier for us to manage bigger numbers of relationships because they act as surrogates for our brains in retaining and collating information on our connections. We don't have to remember other people's favorite movies or their employers because we can look that up when we need it. That's similar to how Google frees our brains from remembering useless information because it's always on hand for when we do cocktail party trivia, and it's also akin to how robots allow us to avoid those dull, dirty, or dangerous tasks. Depending on your perspective, Facebook is either the Google of people we know or the toilet-cleaning robot of our social world.

But with this automation, we lose the human touch. I took my birthday off Facebook years ago because, if people are going to wish me well on my special day, I want to know it's because they actually remembered rather than because a website widget reminded them. Birthday wishes prompted by a computer feel less genuine.

Recent studies support Dunbar's theory. A 2011 Cornell University study found that the number of close confidantes for Americans over the previous twenty-five years had dropped from three to two. A study published in *American Sociological Review* in 2006, before the rise of Facebook, found that socially isolated people—those with zero confidantes—had doubled to 25 percent between 1985 and 2004. Also, the number of people who talked only to family members about important matters had increased from 57 percent to 80 percent over that same time period, while the number who depended totally on a spouse rose from 5 percent to 9 percent.[11] This development probably jibes with what many people are experiencing. We might have hundreds of "friends" online, but in reality we have only a few close confidantes.

The shrinking of our close circles makes for an extraordinarily fascinating trend because it directly counters everything discussed in the previous chapter. It also opposes what Facebook is suggesting. We are definitely putting ourselves out there more than ever, and we are connecting, at least superficially, with more people than before. But at the same time, we're contracting and keeping our most important and personal thoughts and feelings closer at hand. This tendency looks like a counter-reaction to all that sharing and communication. Our desire for privacy is increasing, or perhaps we're hiding our true identities more and more. Friendship is subjective, but closeness usually equates with trust. If more people are confiding in fewer people, it sure looks like we're becoming less trusting. We'll return to these fascinating issues in the next chapter.

## TOO RICH TO FALL IN LOVE

Climbing further up the human relationship ladder, we come to the most intimate rung, our romantic interactions. Here too, as anyone who has a television or a smartphone knows, technological advance is having a profound effect.

We've all seen the statistics on how the majority of marriages end in divorce—52.9 percent in America as of 2011, according to the CDC—but why is that, and is that the full story? A closer look reveals a lot more. The OECD, in its usual understated manner, notes there has been a "noticeable decline" in marriage rates since 1970 in almost all of its member states. The decline has been substantial in countries such as Hungary and Portugal and less so in Denmark and Sweden, where it was already low. But the differences are staggering. In 1970, the OECD average was eight marriages per thousand people, but by 2009 it had fallen to five. In the United States over the same time span, the average dropped from eleven to seven.[12] Also, people are getting married at an older age. Once upon a time, women were ostracized and labeled witches if they weren't hitched by the time they hit fifteen or twenty, but by 2005 the average American woman was twenty-five before tying the knot.[13]

None of this means that people aren't seeking out companionship—quite the opposite, actually. They're choosing to live together longer before getting married, if at all. As the OECD puts it, "cohabitation is [increasingly] used as a stepping stone for marriage or as a stable alternative to it . . . Almost fifty-seven percent of individuals age twenty and older live in a couple household."[14]

Even arranged marriages are declining in some places where they used to hold sway. In Japan, close to three-quarters of unions were arranged in the 1930s, but by the mid-1990s that statistic had reversed almost completely: Nearly 90 percent of people were getting married for "love."[15] Arranged marriages are still prevalent in India, where three-quarters of respondents to a 2013 survey indicated they preferred them to love marriages.[16] The reasons, however, tie closely to the overall issue of why the institution of marriage is declining overall: women's equality. Most societies where arrangements prevail are overwhelmingly patriarchal, and so women generally have lower social standing than men. In India, for example, women face tremendous pressure to ensure their arranged marriages work because of the witch-like stigma that results if they don't, and that's an issue men don't necessarily have to deal with. Similarly, polygamy has been popular in Africa also because of largely patriarchal social structures. It's socially acceptable for men to have multiple relationships and thereby sire more children, but that's not the case for women.

Interestingly, back in the developed world the average length of marriages has stayed largely the same—ten to fifteen years—since the 1970s. People aren't getting married as much as they did, but it's still taking them the same amount of time to get sick of each other.

Some blame these trends on the overt side effects of technology, which is what the media did in 2011 with stories that implicated Facebook in playing a role in one-fifth of all divorces. As the argument went, social networks and the Internet in general have made it easier for extramarital dalliances to happen, just as they've made it possible for people with oddball interests to connect and become friends.

But that argument has arisen with every new advance in communications technology. As Hitched.com editor Steve Cooper put it in a rebuttal to the Facebook divorce stories, "I'm sure at some point during the Stone Age a woman was frustrated because her mate wouldn't step away from the fire and come to bed. More recently, televisions became places of congregation for couples and families."[17] He's right on the money, as a 1966 *New*

*York Times* article proves. In explaining rising divorce rates at the time, the article noted that neglected "working-class wives said their chief rival was the television set."[18] Whatever comes along next—wireless-induced telepathy, perhaps—will face the same scapegoating.

Technology certainly is affecting marriage and divorce trends, but not on the superficial level that some suggest. People used to get together and stay together for a variety of reasons besides love, such as finances, the need or desire to have children, or religious beliefs. Rising prosperity has largely obviated the first reason and helped the second. Changing social mores have contributed too: People who really want children can have them today without having a mate if they're up to the challenge. As for religion, we'll return to that in chapter eight.

If there's any kernel of truth to the idea that online social networks are accelerating the divorce rate, it probably lies in the conventional wisdom that marriage and its delay or demise correlates with an abundance of alternatives. Comedian Chris Rock said it best: "A man is basically as faithful as his options."

## BOOK BURNING AVOIDED

So do we have more options today before or after we get married? I met my wife the old-fashioned way: in a bar while we were both drunk. But that's becoming rarer, according to the statistics. People increasingly are "meeting" online.

In its early days, online dating was perceived a lot like a self-published book: a last resort for losers who couldn't get dates in the real world. It's incredible how quickly that attitude has changed. Of the estimated fifty-four million single people in the United States, a whopping forty million of them have tried online dating. Match.com and eHarmony, two of the biggest sites, boast membership of more than thirty-five million between them, with an estimated one-fifth of all current committed relationships having begun online.[19] Pretty much everyone is dating online now, and it makes sense. Online profiles allow discriminating users to eliminate some of the risk and guesswork and often make it easier to avoid trolls and find the right person. It also nicely opens doors for those people who are more socially awkward or who don't have the confidence needed to approach others out in the wild.

To get an idea of how the dating landscape has changed, I canvassed the best expert I could find: my good friend Dan. He got married in 2004, had a daughter in 2008, and then split from his wife in 2010. In early 2013, he decided to start dating again. Since he was in his late thirties and with a child, he figured he was too old for the bar scene so he investigated a few websites. He settled on OkCupid, run by a company out of New York, because he liked its interface and that it's geared toward casual dating rather than quick flings or serious relationships. His previous adventures in the dating world came before online had gone mainstream, so he's one of those people who has experience in both the 1.0 and 2.0 dating environments.

We met at a café in a comic-book store, the sort of place that is supposedly frequented by stereotypical nerds who can't get dates. He related some of his experiences with online dating, many of which included the sorts of tales everyone has heard by now—back-and-forth electronic chats and messages that led to awkward first meetings in the real world, where it quickly became apparent that one or both parties misrepresented themselves in their profile or photos.

All of that notwithstanding, Dan revealed that he's dating more now than before he was married largely because he can find compatible women more easily. Most dating sites, OkCupid included, require members to fill out surveys providing personality details, likes, dislikes, interests, and so on, which allow users to fine tune their searches and find like-minded individuals, or at least those who possess desirable qualities, or eliminate people with undesirable traits.

One particularly odd question when Dan signed up for OkCupid was whether he preferred flag burning or book burning. The follow-up question asked whether he'd consider dating someone who answered the opposite. "At least you can tell early if someone is into book-burning, instead of finding that out eventually," he says. "Once you've done it for a while you start looking at people's profiles and you understand if they're annoying or not, or if you have a lot in common."

Dan may be dating more, but getting an empirical measure of whether the population as a whole is dating more is impossible, for now, mainly because the concept of dating itself is still relatively new. It's also somewhat rare, occurring mostly in prosperous Western countries with relative gender equality where women are free to choose their mates.

We can charitably trace dating's origins to the eighteenth and nineteenth centuries, when young women entertained gentlemen callers under the watchful eye of a chaperone, which of course brings up images of Jane Austen novels. But it didn't really start in what we recognize as its current form until the early twentieth century, when couples went out to a neutral location such as a movie theater or restaurant. The "going out" part generally came from individuals not wanting their potential mates to see their usually squalid homes and not wanting to court under the watchful, disapproving eyes of parents or guardians. Dating took off as a social phenomenon after the sexual revolution of the 1960s, when college students of different genders lived in close proximity in dormitories. Increasing gender equality and rebellion against norms—plus freer access to booze, drugs, and contraception—got it all rolling.

Now it's unquestionably technology that's accelerating social changes. My friend Dan has observed a marked uptick in the number of people he's dating, and statistics also show that the speed of relationships is increasing. The average length of courtships that started online and resulted in marriage is about eighteen months, while those who met offline come in closer to forty-two months.[20] It seems that the more people know about each other up front, the faster their relationship—which may or may not ultimately involve marriage—is going to progress. On the other hand, as the other numbers show, this acceleration isn't leading to an increase in marriages.

Our online dating habits will become increasingly clearer as the data set expands. Years from now we'll have a much better idea of how technology is affecting courtship. With the information we have now, though, it looks like the multitude of choices is indeed making it harder for many people to choose. Even back in 2003, a *New York Times* article illustrated the issue: "Of the one-hundred-and-twenty men she traded messages with online in her first four months of Internet dating, Kristen Costello, thirty-three, talked to twenty on the telephone at least once and met eleven in person. Of those, Ms. Costello dated four several times before realizing she had not found 'the one.'"[21] If the declining marriage numbers and the increase in age at which people are getting married offer any indication, we do appear to be suffering from the paradox of choice when it comes to selecting our mates. Dating online increases the pool of those available mates, so dating more people takes longer, which may contribute to driving up the marriage age.

## THE EQUALITY OF CHEATING

So what's happening on the other side of dating, the extramarital affairs that often lead to divorce? With technology providing access to Chris Rock's proverbial "options," are affairs increasing in frequency as well?

Noel Biderman, founder of the affair-facilitating website AshleyMadison.com, has some thoughts on the issue. The site, based in Toronto, operates in more than two dozen countries and has a floating membership of around nineteen million. Biderman is something of a pariah, a bogeyman that the public and the press have blamed since the website's launch in 2002 for aiding and abetting infidelity and the destruction of families that often follows. Yet he himself is happily married with children—"Divorce lawyers don't have to be divorced, do they?" he quips. It's clear he's used that line before.

He's dressed casually, wearing a comfortable-yet-snazzy hoodie that makes him look like a slightly more distinguished Mark Zuckerberg. He's sporting rainbow-colored socks, visible just under his desk, which indicate a certain sense of playfulness. He peppers our conversation with the odd joke, so he clearly doesn't take himself too seriously.[22]

Biderman talks fast and seems to have an answer for everything—except when I ask if his website is leading to more affairs happening than ever before. "The problem is, we don't really have a good historic sense," he says. "How unfaithful were we in 1972? I don't know and I don't think anyone really knows." He makes a good point. Humans have been having affairs for centuries, but nobody has been tracking them until the likes of AshleyMadison came along. "It's only now that we're getting a sense of the volume of infidelity," he adds.

The data that his website is gathering on its users provides a gold mine for social scientists, and Biderman is partnering with universities in Michigan and California to study it. A large monitor hanging on the wall behind his desk displays some of AshleyMadison's most pertinent information. At a glance, we can see that 6,447 new sign-ups have taken place so far today, about 7 percent above the average, while average revenue per user is down slightly on the day by 2 percent. About twenty thousand affair seekers will sign up before the day ends. It's like watching a stock market, except the tickers monitor the statistical ups and downs of people cheating on each other.

I ask whether he considers the tracking and sharing of such data with scientists, who can study what it means, to be the socially constructive "good guy" part of what he does. "No one's ever put it that way, but yeah, I guess it is," he says. "Part of the legacy we'll leave behind is this data set, which people much smarter than me will dive into. With that information, people might approach marriage differently or better. I don't know?"

But Biderman is well rehearsed in answering the charge of whether AshleyMadison is a social ill overall or merely a vehicle for affairs already happening in the first place. He maintains that he's just an innocent guy out to make a buck from a long-standing reality of humanity. "If an affair is about meeting someone and not getting caught, I just need to cannibalize that," he says. "I don't have to generate the demand, it already exists because of our genetics. I can't [make it happen] with a thirty-second TV commercial. No one is that pliable."

He is seeing a change in who instigates affairs, however. That role historically has gone to men, but more women are doing so because of improved equality overall. In this way, technology is acting as an accelerant because it provides women with the constructs through which affairs can happen. For years men have had their own set of brick-and-mortar platforms—strip joints and massage parlors, for example—but women haven't had those equivalents. Websites such as AshleyMadison and perhaps even Facebook are filling that void. "Technology has not changed the male infidelity landscape," Biderman says. "It *has* allowed for the cannibalization of the more traditional environments where it has happened, and I believe it's currently impacting the female infidelity landscape."

The majority of affairs are now catalyzing online as opposed to at work or otherwise "in real life," according to the company's statistics. The typical affair is also accelerating in speed. Whereas in the past it might have taken weeks or months for an affair to come together, now it typically happens in thirty-six hours: Users sign up to AshleyMadison in the morning before work, hunt for willing partners and send them messages during their lunch hour, respond to messages and set up meetings the following morning, then meet after work that day. Before, people who engaged in affairs had to concoct elaborate charades to make that secret phone call—"Uh, I have to go walk the dog"—whereas now they're sending and checking messages with their spouse in the same room.

According to the stats monitor, 62 percent of women are checking their inboxes during prime-time evening hours, possibly while sitting across from their husbands on the couch.

Smartphone GPS functions only accelerate the situation more. Apps geared toward enabling affairs can tell users if there's someone else in the same hotel who's raring to go, for example. "The technology has shortened the windows," Biderman says. "There is that time consolidation for sure."

It's easy to conclude that technology is accelerating affairs and therefore leading to an increase in their numbers. The statistics aren't conclusive, but surely this development plays a role in rising divorce rates. Maybe those Facebook divorce stories were correct after all? Biderman isn't sure, but he does believe he knows the underlying reason for most affairs. "I've never seen a study that says the longer you're with someone, the more sexually attractive they are to you."

## NOT TONIGHT, I HAVE AN E-MAIL

Which brings us to the most basic of human relationships: sex. It's the only biological relationship that we humans *must* have, otherwise we die out. Everything else—dating, courtship, marriage, affairs—forms a jumble of social constructs centered around this basic genetic imperative.

Shrinking marriage rates, rising divorce rates, accelerating dating, and faster affairs—plus what some call the Internet's "pornification" of media and culture—surely add up to humans becoming one big, sex-crazed lot, right? Just as technological advance is causing exponential growth in the number of transistors that can fit on a microchip, it's also exponentially affecting and breaking down those social constructs, inevitably leaving us with only our base biological urges, right?

Well, maybe. Sex is also hard to quantify, given that we've been doing it—literally—for as long as we've been on Earth. But are we having more sex today than Blarg the caveman did in his day? Or the Romans did during their orgies? Probably not, but it's impossible to tell definitively one way or the other.

Another, more specific problem with measurement lies with the changing definitions of sex. In a recent study by the Kinsey Institute, which surveys all things sexual, nearly half of respondents considered manual genital stimulation as "having sex," while 70 percent said the same about oral and 80 percent said so about anal.[23] Sex is like a snowflake or *2001: A Space Odyssey*—no two versions or interpretations of it are alike.

Also impeding the study of sex is that most data comes from interviews with individuals who have all manner of reasons to distort, skew, and withhold information, and otherwise lie about their exploits even to unassuming and nonjudgmental researchers. As a result, many social scientists who study the subject try to level their findings by sticking to a simple rule of thumb: Men tend to over-report and women often under-report.

Sex could be on the rise because of lengthening life expectancy, better health as people age, and other technological advancements—a little blue pill comes to mind—all of which make it easier for older people to engage in what have historically been considered more youthful pursuits. On the other hand, the increasing pressures and stresses of daily life plus the multitude of other pleasure options are likely countering those factors. Technological and economical advancements may actually be cancelling themselves out.

Sex is a little easier to measure if we consider only the modern era. In that case, yes, a definite rise happened through most of the twentieth century, and technology was to thank or to blame. (Your mileage may vary.) In 1900, only about 6 percent of unwed teenage females had engaged in premarital sex, but that number skyrocketed to about three-quarters by the end of the millennium. We can attribute the low incidence a century ago entirely to the near certain chance—72 percent—of pregnancy. But ahoy there: along came the mass-produced latex condom in the 1920s, and then several decades later the birth control pill. Between 1960 and 1964, those delightful new technological developments were used respectively in about 22 percent and 4 percent of premarital dalliances. By 2002 their usage had risen to 51 percent and 16 percent, respectively. There's no doubt: Contraception was key to the sexual revolution. People didn't have sex for pleasure all that much before then simply because it was too risky. As a 2010 study of the phenomenon put it, the "rocket-like rise" in premarital sex and subsequent effects on marriage and divorce were due almost entirely to these technological changes:

> *This is traced here to the dramatic decline in the expected cost of premarital sex, due to technological improvement in contraceptives and their increased availability . . . First, individuals weigh the cost and benefit of coitus when engaging in premarital sexual activity. Second, they associate with individuals who share their own proclivities. Such a model mimics well the observed rise in premarital sexual activity, given the observed decline in the risk of sex . . . Improvement in contraceptive technology may also partially explain the decline in the fraction of life spent married for a female from 0.88 in 1950 to 0.60 in 1995. This is due to delays in first marriages and remarriages and a rise in divorce. Historically, the institution of marriage was a mechanism to have safe sex, among other things. As sex became safer, the need for marriage declined on this account.*[24]

Our question now becomes whether additional technological advances have continued to lead to more sex. The data here is considerably less convincing.

Some observers, such as writer Pamela Paul, have argued that technology's effects on media dissemination, especially via the Internet, have led to a "pornification" of culture in many Western countries. On the Internet, hard-core porn lies only a couple of clicks away. The medium has gone mainstream, leading to a general degradation of social values and a desensitization to the negative effects and aspects of pornography, which can include objectification of women and violence toward them. As Diane Abbott, British Labour party politician, puts it:

> *The rising numbers of girls having under-age sex is alarming. It is not a cost-free phenomenon. It poses public health policy challenges and social challenges. The underlying cause must be the "pornification" of British culture and the increasing sexualization of pre-adolescent girls . . . Too many young girls are absorbing from the popular culture around them that they only have value as sex objects. Inevitably they act this notion out. Government needs to respond to spiraling under-age sex, not with pointless schemes to teach abstinence, but with better . . . teaching in schools for both girls and boys.*[25]

Such arguments seem logical, but no solid data underlies them. In the United States, the Center for Disease Control reports that in 1988 the average American man had five or six sexual partners in his lifetime, while the average woman had one or two. Those numbers haven't grown much over the past few decades, with men now averaging 6.1 and women 3.6.

Bryant Paul (no relation to Pamela Paul), associate professor at Indiana University and affiliated scholar at the Kinsey Institute, considers the "pornification" argument nonsense. "The sexualization of youth is not something that's very new," he says.[26] Sex is an ongoing phenomenon that ebbs and flows with the times and cultural changes. As uncomfortable as the thought may be to many people today, he says, sexual relationships between grown men and boys were cherished in ancient Greece and Rome, for example.

Evidence also suggests that technological advancement actually stifles sexual activity. In sub-Saharan Africa, men often have three or more sexual relationships happening *concurrently*, which puts to shame that six-person lifetime average for American men. Justin Garcia, also an assistant professor at Indiana University and research scientist at the Kinsey Institute, says the Nagunda farmers and foragers in Africa have "way more sex" than people in developed nations. "It's viewed as baby-making work," he says. "It's regimented. You have sex three times in a night, then you take four days off."

If there is a pornification problem, it might have started with the Industrial Revolution. Prior to the eighteenth century, children were often present when their parents had sex, primarily because families shared bedrooms. They also witnessed childbirth, which usually happened at home. But then "adolescence" was invented, at which point sex education took place when we were older and deemed responsible enough to handle it. Now, the most readily available place where kids can learn about sex is through Internet pornography. If their views and subsequent actions are skewed, it's hardly a surprise. "They need to be exposed to some of this stuff," says Paul. "They don't have to watch hard-core pornography, but they need to know where babies come from and what sex is."

Pornification may be happening, according to the data, in the kinds of sex people are having. In 1990, only half of American men and a quarter of women reported being regularly on the receiving end of oral sex, a number that rose to 80 percent for both genders by 2006.[27] The same goes for anal sex: Only a third of women reported to having tried it in 1992, compared to nearly half by 2010.[28] But people rarely tell the truth about their sexual escapades, so these numbers merely may reflect an increased honesty. If that's the case, it's still important because it indicates significant changes in social mores.

Some of those changes in social customs may come from exposure to pornography, which could be inspiring people to take sexual evolution another step forward. We used to have sex mainly to have children; then it was primarily for pleasure. Now, sex may be becoming a sort of performance art. Just as our growing creation and dissemination of photos, status updates, and other media can be considered a form of public performance, so too can our growing liberalization and experimentation between the sheets. If we can let it all hang out online, why not in the bedroom?

Technological, economic, and gender equality advances spurred sexual activity during much of the twentieth century, and it's likely that all of these factors are continuing to contribute to its evolution. But it's not clear that we are all becoming sex maniacs. In a global and historical perspective, people in developed countries today are downright prudes who are returning to a sexual level last seen in the sixties, before the chill that HIV and other sexually transmitted diseases brought on in the eighties and beyond. But even if that's the case, oral and anal sex were accepted widely in ancient Greece and Rome, and much of the developing world has a lot more sex than people in Europe or North America do. It may be impossible to enumerate the entire human history of sex, but it's not out of bounds to suggest that technological advances are resulting in *less* sex than ever before.

## THE X FACTOR IRL

Just as the world is currently experiencing an economic paradox—where the Gini coefficient of equality is narrowing between countries, yet expanding within many nations—so too is there a contradiction in human communications and relations. As the previous chapter showed, the rate at which we're communicating with one another and expressing ourselves is growing dramatically. But as much of this chapter illustrates, we paradoxically are shrinking inward. The world is inching toward greater harmony and we are making many more "friends" along the way. But the real data on friends, love, and sex indicate that those expanded relationships are becoming increasingly superficial and that the deep ones are declining.

Techno-pollyannas focus only on the positive and potential effects that ubiquitous connectivity allows, but they shy away from some of these very real demographic results or rationalize them away as misinterpretations of the data. Worse still, some suggest that even *more* technology might be the solution. Witness the proliferation of social-media management apps such as Hootsuite or Tweetdeck, which allow users to follow even more online conversations by grouping them into categories supposedly easier to track.

Sherry Turkle, a professor at MIT and author of *Alone Together*, has reached many of these conclusions, wherein technology has brought about a lot of unintended consequences as far as relationships are concerned. She also thinks the addition of more technology to the problem is absurd. Communications advances were intended to improve the way we do business, but they weren't supposed to change who we fundamentally are. But that's what has happened. Families and relationships have all changed dramatically over the past century. "We stopped talking to each other because texting is simpler and it allows you to avoid some of the difficult parts of talking," she says. "Connectivity left us seduced by the illusion of companionship without the demands of friendship. It left us seduced with connection instead of conversation."[29]

Much of the problem, she elaborates, is that the virtual conversations so many people have today don't align with reality. When you have an "in real life" conversation with someone, you present yourself how you really are rather than how you want to be perceived. Nor can you control what happens, as you often can with technological conversations. In virtual conversations, for example, you can answer an angry text message hours later, a luxury you don't have when someone is literally shouting in your face. Virtual conversations in all their many forms remove these "X" factors, some of which are vital to forming and deepening relationships. "You're taking people out of an equation where people have always been and in the course of having them there, attachments and relationships developed," Turkle says. "It's faking the conversation, in a certain sense."

Perhaps these recent developments aren't irreversible, nor might they all be temporary, like a technological hangover that many people are experiencing in the wake of the digital-Internet diffusion of the past two decades. Turkle agrees, and she's optimistic about the future. She thinks young people understand the nuances much better than some of us older folks. Growing up with ubiquitous communication rather than flocking to it because of its novelty arms them with better coping capabilities. Most kids know, for example, that hanging out on Facebook at three in the morning isn't nearly as fun as doing so in real life. These are lessons that the rest of us, hopefully, will learn soon. "Technology can make us forget what we know about life and for a couple years there, that seemed like a good idea," Turkle says. "We're ready to take a deep breath and do it in a way that meets our goals as people a little bit better."

# 7

# Identity: God Is the Machine

*In the kingdom of glass everything is transparent, and there is no place to hide a dark heart.*

—VERA NAZARIAN

"If you have something that you don't want anyone to know, maybe you shouldn't be doing it in the first place." That was Eric Schmidt's response to a question about whether Internet users should share information with Google as if it were a "trusted friend." Critics quickly pounced on his nonchalant reply, which aired in a 2009 NBC News special, pointing out that the company's chief executive just a few years earlier had blacklisted certain news organizations for publishing details about his salary, home neighborhood, hobbies, and political donations, all gleaned from Google searches of course.

Hypocrisy aside, the truth will get out. It always has a way of becoming known. But at the same time, there's a big difference between keeping secrets under wraps and not creating them in the first place. If, for example, I pulled off the perfect bank heist, no one would know about it unless I spilled the beans. What Schmidt was suggesting, however, is that I shouldn't have robbed the bank in the first place. He's right, obviously, because doing so would be criminal and the repercussions dire if anyone found out. But what if criminality devolves into something that's merely embarrassing, like secretly enjoying Justin Bieber's music? There's certainly nothing wrong with it per se, but it does have social ramifications. Was Schmidt suggesting we shouldn't entertain guilty pleasures at all? Should Google and other online services be able to expose all of our secrets, no matter how trivial?

These questions intensified in volume and severity a few years later when the focus shifted from corporate snooping to government surveillance. In 2013, former National Security Agency contractor Edward Snowden began his symphony of whistle-blowing revelations on how the US government was spying on ordinary citizens around the world under the guise of rooting out terrorists. Using the same rationalization as Schmidt—that regular law-abiding people had nothing to fear—the NSA spent the better part of a decade building a comprehensive surveillance machine that ultimately got its figurative fingers into every conceivable digital crevice. The revelations came fast and furious: the ability to listen in on more than a billion phone calls a day, gathering thousands of e-mails from people not connected in any way to terrorism, spying on (even friendly) foreign diplomats, engineering back doors into the servers of big Internet companies, agency employees using the system to spy on ex-girlfriends, and creating systems that can break any form of digital encryption. The only question remaining at the end of 2013 was whether any lines remained that the NSA *hadn't* crossed.

These two examples highlight what has become *the* issue of the digital age: privacy. To start, there's the question of expectations. In an era where so much of our personal information lies at anyone's fingertips—much of it put there voluntarily by us ourselves—how much of a right do people even have to privacy? There's also the question of whether that expectation should change based on our social and economic standing. Why do wealthy CEOs feel it's okay to expect their information to remain private? Then there's the fundamental question of that imminent trade-off: Do we really need to sacrifice some or all of our privacy in the name of security?

The most important part revolves around why the concept of privacy exists. It gives a space in which we can shape our identities. Privacy very much molds us as collective and as individual beings: who we consider ourselves to be and how we conduct ourselves in relation to the people around us. Technology may be making us more economically comfortable, helping us live longer, broadening our potential for relationships and personal expression, and creating new avenues of employment, but with its inextricable links to privacy it's also subtly and fundamentally shaping who we are. What happens when we don't have that identity-building space anymore?

## TOP SECRET LICENSE PLATES

First we need to gauge what privacy is, how much we have, and whether we possess more or less of it than before. Unfortunately, as with sex, defining it isn't easy. It's another snowflake game where no two definitions agree.

Samuel Warren and Louis Brandeis, two lawyers, suggested a simple definition of privacy back in 1890 in response to recent technological advances in journalism and photography. They called it "the right to be left alone." Academic Alan Westin took the idea further in 1967 by describing privacy as the ability of an individual to decide what information is communicated about him or herself when and how. Social psychologist Irwin Altman came up with the privacy regulation theory in 1975, which holds that privacy is a state of social withdrawal that people sometimes desire. Since then, privacy has become an active conflict between an individual's right to secrecy and society's right to maintain order and balance. Whichever way it's considered, it's an important ingredient for group preservation and a stable social system.

All of the definitions are correct, which makes privacy, like a cut diamond, a multifaceted subject. Further complicating the matter is that the study of it is multidisciplinary; it's equal parts law, psychology, sociology, and technology. To understand and define it completely, top experts from each of those fields would need to come together and pool their efforts. That kind of united undertaking has happened before in other areas but generally only under exceptional circumstances. The Manhattan Project, which allowed for the creation of the atomic bomb in the 1940s, represents such a multidisciplinary effort. We may never properly understand privacy and its effects until a similarly wide-ranging project takes place, and that might not happen unless some sort of crisis or catastrophe necessitates it. But with increasingly frequent revelations about privacy breaches and governments spying on their citizens, we're steamrolling toward that event horizon.

In the meantime, some people are trying valiantly to understand and measure privacy, whatever that means. Many of them live in Canada, where privacy seems to have become a national obsession, right up there with hockey, beer, and poutine. Perhaps it's the brutally cold winters, which force us to bundle up and hibernate, or the drastic contrast between our traditional low density urban and rural spaces and the increasingly intense urbanization in our biggest cities that make us so concerned about our personal space. Or maybe we're just exceptionally paranoid. Whichever it is, an awful lot of my fellow Canadians are working on studying and protecting privacy. As it turns out, their findings and policies have a habit of going global. You could call privacy one of Canada's biggest exports, right up there with Justin Bieber, Cirque du Soleil, and maple syrup.

Tracy Ann Kosa is one of those privacy-obsessed individuals. She works at Microsoft's Trustworthy Computing division in Redmond, Washington, but she began her career putting together privacy impact assessments (PIAs) in Canada, mainly for the public sector. The process, she says, was tedious and expensive; just the guide on how to conduct one ran to three hundred pages. She began by describing an organization's product, service, or system, followed by an explanation of the business processes that support it, supplemented by diagrams of how data flowed through it. Then came an assessment of how it worked in conjunction with ten established regulatory privacy principles, as well as the legal requirements.

If it sounds boring, it was. Kosa did it for ten years and decided there had to be a better way. Even at the end of completing a PIA, meaningful documents were rare because the process itself wasn't standardized. "We'd institute all these rules on an organization with the notion that we're now protecting privacy, but when we actually look at what constitutes informational privacy and what is possible, the two things aren't really connected," she says.[1]

In 2011, she wrote a paper with several colleagues that proposed a streamlined way for organizations to measure the privacy impacts of whatever they were looking to build.[2] They outlined nine measurable states for products, services, or systems:

1. totally private, where the user's existence is unknown to the provider;
2. unidentified, where the user's existence is known, but the individual is not;
3. anonymous, where some information about the individual is known;
4. masked, where the user can be known, but the links to his or her identity are hidden;
5. de-identified, where an individual's identity is not known, but can be when the known information is linked to others;
6. pseudonymous, where identity is known, but incorrectly;
7. confidential, where identity is known but only in a defined setting;
8. identified; and
9. totally open and public.

Kosa and her colleagues posited that a number of factors govern how and why people fit into those different states at different times. People more likely fall into the open end of the spectrum, for example, at a dinner party with people they know. They clamp down, however, at a police station or if they know they're being recorded. Other factors that determine how much information people share include the social status of the so-called privacy invader, the presence of authority figures, existing relationships between involved parties, and the expected audience for the given information. The authors also settled on four types of disclosure: information that people voluntarily share about themselves, what they disclose about their property or objects, and then third parties' disclosure about each of those two.

The model that Kosa and her colleagues proposed would run in the background of a given process, such as sending an e-mail, and instantly calculate all of those factors and possibilities, then display for the user the likely privacy state of the action. An e-mail program could, for example, tell a user that his message would likely qualify for the "confidential" state, where the information in it and the sender's identity are secret, but could be easily determined by an interceptor. Armed with that information, the user could then determine whether to proceed.

Such a system, if it worked well and was applied universally, would give people a good way to measure how private their various technology usage was. Is Gmail more secure than Skype? Is Amazon more private than eBay? A universal measurement system would tell us and could kick off a privacy arms race. A good portion of users likely would come to value more secure services, which would result in providers beefing up their privacy, rather than ratcheting it down as many have continued to do. Some systems would be designed with up-front limitations, where the provider discloses to the public that, for example, "it's just not going to get a level higher than four," Kosa says.

The problem with the idea was that, if implemented, it could be really annoying. Microsoft tried something in this vein in 2006 with its Windows Vista operating system, which displayed messages to users nearly any time they tried to do anything. "Are you sure you want to copy and paste that?" "Are you sure you want to close that window?" "Are you sure you're sure?" "Are you sure you're sure you're sure?" The constant stream of warnings, while handy in preventing unwanted events from occurring, helped kill the operating system. People weren't seeing the benefits, but they definitely felt the frustrations with the supposed safeguards. Giving users a constant stream of security ratings that they have to accept could be just as undesirable. "It was not a well-received notion, to put it mildly," Kosa laughs.

Product and service makers and even watchdogs still maintain a general reluctance to agree on a unified field of privacy, she says. As long as that resistance persists, efforts to measure and standardize privacy settings and protections will continue to face insurmountable obstacles.

The effort to quantify and create a hierarchy of privacy states isn't all that unusual. The military has been doing it for ages. Various documents get different grades of security: classified, top secret, ultra, and so on. John Weigelt, national technology officer at Microsoft Canada, understands that system better than most. He started his career in the military, where one of the first systems he helmed was personnel records for the Defense Department. That information is considered extremely secret so protecting it was of the highest importance, much more so than simpler information like administrative budgets, for example. The general public, however, hasn't yet arrived at those levels of distinctions. "We haven't come so far in the privacy world," Weigelt says. "We tend to treat a license plate the same as a social insurance number. We don't have this idea of gradations."[3]

## PARENTS, THE REAL ENEMY

Like dating and marriage, privacy is a social construct. There's no biological reason for it. Blarg and our other forebears had no concept of it. They slept together huddled in caves, where everyone knew what everyone else was doing all of the time. They wore animal furs only to keep warm, not necessarily to hide their cave jewels from one another.

Only with the advent of the Industrial Revolution did modern notions and expectations of privacy for common people emerge. Just as electricity and heating allowed for families to live in separate rooms, so too did they create private spaces in homes where individuals could escape their children, parents, siblings, or spouses. Household secrets then had the opportunity to germinate. Privacy is largely a technological byproduct, powered at first by electricity. Additional developments, such as the postal system, the telephone, personal banking, and medical institutions, further enshrined the public's expectations of protections on personal space and secrets. Laws followed in each instance.

It's possible that Kosa and other researchers are barking up the wrong tree, though. Privacy may not be all that hard to measure after all, because historically it shows strong links to income levels. As Schmidt's example indicates, rich people generally have more opportunities to acquire or maintain privacy. Big-shot execs and celebrities hounded by paparazzi buy mansions and estates with twelve-foot walls, vicious dogs, alarm systems, and security guards. Google and Facebook urge their rank-and-file users to disclose more and more of their personal information, but their CEOs ironically share very little and get upset when someone digs up such details (Mark Zuckerberg famously said that the age of privacy is over—just not for him. He keeps his Facebook profile locked down). This historical dichotomy applies on a global level too. In poorer parts of the world, families still live closely together, which means that individuals still know what their relatives are doing most of the time. In the end, privacy is an intangible commodity. You can have as much of it as you want as long as you value it and can pay for it. It may not be possible yet to measure privacy on a fine-grained scale, but we can draw broad strokes between it and how much money a person has.

Today, many people take it for granted. For a period of time in the twentieth century, it was a sacred cow, enshrined in law and architecture, an inalienable and infallible right. But new technology has ushered in a new reality. Consider the disconnect between how authorities treat different technologies. Law enforcement agencies and even Internet service providers think it's perfectly acceptable to tap into our Internet traffic to see what's going on inside it, but it's illegal for them to listen to a phone conversation without a warrant. Much of the world is stumbling from this old, comfortable assumption of privacy to the public default of a new, unsettled Wild West.

The transition has created a shift in values. Following the laws of supply and demand, people value commodities less when they're plentiful and readily available, but they crave them more when they become rarer and harder to acquire. That's what's going on with privacy, according to Jennifer Stoddart, who was Canada's top watchdog on the subject between 2003 and 2013.

Stoddart is something of a celebrity among the privacy crowd (Kosa admits to being "a total fangirl"). In Stoddart's tenure as Canada's privacy commissioner, she took on the likes of Google and Facebook and won. In 2007, she acquainted Google with a part of Canadian law that makes it illegal to publish a person's photo without his or her consent, except for journalistic or artistic purposes. The company had to blur faces as well as license plates and other potentially revealing details in its Street View navigation service. In 2008, Stoddart became the first privacy regulator in the world to challenge Facebook on how it allowed third-party developers to access users' data without their consent, which again is illegal in Canada. The website initially agreed to change its system so that users had to opt in to allow their photos, videos, and personal information to be shared, but then it backtracked in 2009 in what looked like a challenge to the commissioner's authority. The company ultimately decided against going to war and instituted the requested changes in 2010. Both Google and Facebook altered their services on a global basis, deciding that it was better to keep them uniform and avoid similar challenges in other countries. But that wasn't enough for Stoddart, who in 2013 began lobbying the Canadian government for the ability to institute eight-figure fines for corporate privacy violations.

Depending on your perspective, Stoddart is either a maverick or a bulldog, and her influence has become so strong that Internet companies have tried to discredit her behind the scenes. I've heard lobbyists suggest that she was a political opportunist who was trying to buff her public profile or that she was a hypocrite for not going after the government and its privacy violations. On the latter, they had a point, since Canadian spy agencies have been willing and complicit partners of the NSA, according to Edward Snowden's leaked documents. Speaking with Stoddart, it's clear there's no love lost on her part either. "To some extent, it's a culture of impunity among that community," she says of the Internet companies. "You just go ahead and do something that's cool and will sell and you don't look as to, 'Are there any rules that have been established to guide our use of personal information in this context?'"[4]

I asked her if we're more or less private today than we used to be as a result of all these new Internet services and capabilities. She gave a paradoxical answer: "Both." We aren't sharing caves and running around naked anymore, but we certainly aren't skipping blithely through the golden age of privacy—and that's a shift that was happening before the Internet came along, she explains. The spread of other media, including movies and television, effected as much social change throughout the twentieth century, which means we've been getting increasingly liberal for some time.

She mentions a story in the news about the twisted sex lives of penguins. Antarctic explorers witnessed "astonishing depravity" back in 1910, such as male penguins trying to mate with dead females, but they didn't dare report it at the time because it was just too shocking. Their observations only came to light a century later.[5] "People will talk about things now that a generation ago you wouldn't talk about, much less even three years ago," Stoddart says. "It was taboo even to publish this in Edwardian England. That's just penguins, but it's a parallel development."

She's not sure how to measure privacy either, but she's convinced of two points: We have less of it today thanks to technological intrusion and sharing, whether voluntary or not, and people value it more. In a study conducted by her office, two-thirds of respondents said that protecting their information would be one of the most important issues in the coming years. Many of those responses, to her surprise, came from young people, a group generally thought to value privacy the least. "The very demographics that some old folks like me think have no sense of privacy do in fact, when they're asked, have a very strong sense of privacy," she says. "It's just different from the older generation."

Ian Kerr, Canada Research Chair in Ethics, Law, and Technology at the University of Ottawa, couldn't agree more. In 2007, he conducted a panel study with Canada's various provincial privacy commissioners and a group of kids. One of the commissioners, all smug and adult-like, told one of the young participants that his communications over the likes of BlackBerry Messenger and MSN were in fact owned by the respective service providers. The youngster said he knew that, but he didn't care. He wasn't worried about a faceless corporation knowing what he was texting because it was more private than talking on his phone, which his parents— his real concern—could overhear. "It was a real eye-opener because it told the commissioners that there wasn't a lack of privacy value," Kerr says. "It's that young people place their salience in a different location than other people would."[6]

In another example, one youth was asked why he was on Facebook despite hating it. He said he had to be on it because if he wasn't other kids at school would make a fake account in his name and say things about him that weren't true. To Kerr, the anecdote proves that younger people care deeply about privacy and what it means to their identities despite all the information they voluntarily share on the Internet. "Here was a kid who was pre-emptively protecting himself against a kind of identity theft," he says.

A 2013 study of social media and privacy by the Pew Internet Project found that nearly two-thirds of teen Facebook users keep their profiles private, while most reported high levels of confidence in their ability to manage privacy settings. More than half reported sharing inside jokes or cloaking their messages in code in some way and more than a quarter posted false information, like a fake name or age, to help protect their privacy. Some of the responses and thoughts on privacy amusingly mirrored my own teenaged discoveries on sharing secrets. "I have privacy settings, I just don't really use them because I don't post anything that I find private," said one teen. "I feel like I kind of just have a filter in my brain. I just know that's not a good idea [to post revealing content]," said another.[7] Older people may instinctively think that the younger generation doesn't value privacy, but the studies show that's hardly the case.

The problem, increasingly, isn't about what we're sharing voluntarily, it's the information that's being gathered, sorted, processed, and analyzed without our knowledge or consent. Here privacy is paradoxically declining in quantity and increasing in value. Whether it's Google and Facebook sucking up an ever-larger amount of our personal information, an airline tracking our travel habits, or IKEA asking for our zip code while we're paying for that bookcase, companies are learning that there's far more value in knowing who we are, where we are, and what we do than in simply taking our money. Having all that information means they can take our money more effectively and more frequently in the future.

When it comes to the government, the implications are more dire. Governments can and will gather information about their citizens to understand them, sometimes in collusion with those companies, and in order to classify them into groups that can be treated differently. This isn't just some vague dystopian nightmare. It's already happening. Just ask anyone profiled by religion or ethnicity and put erroneously on a "no-fly" list. In 2010, Canada scrapped its mandatory long-form census because, ministers claimed, it was invading people's privacy. Yet revelations continue to expose that governments are building huge data-monitoring capabilities and are then using those facilities to spy on their citizenry with impunity. The idea of privacy has been flipped on its head. People don't have to disclose their own information voluntarily anymore; it's being taken from them regardless of their wishes.

It's difficult to predict how the future of online privacy will unfold, but money will continue to hover at the center of the equation. Private encryption will become easier and paid alternatives to Google and Facebook that don't collect users' information—imagine each Internet search costing a penny—are likely to arise. But again, much like celebrities' walls and guard dogs, such options will be available only to people willing and able to pay for them. The rest of the public may be in for a bumpy ride. "That's disproportionately going to affect poorer people, who will be treated in particular kinds of ways based on the social categories that they're put into," Kerr says. It's not so much Orwell's Big Brother as it is Kafka's trial of Joseph Kay, where a dossier of information accrues on an individual, who is then put into certain categories without his knowledge or consent. "It's potentially affecting your life's chances and opportunities."

## EVERYONE ENTERS, NO ONE LEAVES

Privacy isn't just about secretly listening to Justin Bieber songs. As decades of research have shown, the amount of privacy people have shapes their behavior and who ultimately they are. Surveillance or even the specter of it can cause people to act differently than they normally would—in most cases, more conservatively, guardedly, and in accordance with laws or established social mores.

A 2005 British Home Office study of London's pervasive security cameras, for one, found significant reductions in premeditated crimes in most places where Big Brother's eyes were watching.[8] Similarly, a 2008 study on the effects of cameras in Swedish soccer stadiums found unruly behavior to be two-thirds lower when they were present than when they weren't.[9] A 2011 Australian study found that people were more likely to condemn the socially "bad" behavior of others if they felt they were being watched. As the researchers behind the study wrote:

*[We] presented students at the Campus Universitaire de Jussieu in Paris with stories of two moral transgressions: keeping the money found in a lost wallet and faking a résumé. For some participants, the scenarios were accompanied by an image of a pair of eyes, for others the scenarios were accompanied by an image of flowers. Those given the version with the eyes rated the actions as less morally acceptable than those who saw the flowers.*[10]

Ron Deibert, another Canadian, takes a particular interest in privacy and its implications on identity. He's an author, professor at the University of Toronto, and founder of the Citizen Lab, a group that monitors Internet surveillance and the human rights abuses it sometimes causes. He mentions Iran's Green Movement in 2009 and 2010 as an example of how the government used the Internet to stifle dissent. Social media initially enabled opposition forces, but authorities eventually learned the ropes and used the same tools—Facebook and Twitter—to entrap and immobilize protestors, creating a climate of fear. "They were able to convincingly convey that all of the communications were being monitored," Deibert says. "The Iranian Green movement quickly disaggregated into splinter cells and people worried about who was talking to who."[11]

The negative effects don't even have to start with technology. As Iron Curtain chronicler Anna Funder writes, life under the watchful eye of the Stasi in East Berlin was stifling and oppressive:

*It was inconceivable that a person would ask a stranger, a total stranger whether they lived near the border. It was also inconceivable that the stranger would ask you whether you were thinking of escaping . . . Relations between people were conditioned by the fact that one or the other of you could be one of them. Everyone suspected everyone else and the mistrust this bred was the foundation of social existence.*[12]

The effects, great or small, can range from individuals omitting certain flaws or features from their online networking or dating profiles to entire countries outlawing certain modes of thought. The knowledge or even suspicion that we're being watched pushes us further from freedom in thought and action. On one hand that's good because it preserves the social order, but on the other it can change significantly who we are or want to be. "Left unchecked, surveillance can create a climate of self-censorship," Deibert says. "If they know they're being watched and all of their activities are being monitored, people tend to be more conservative. That's something we have to be conscientious of."

In 1787, English philosopher and social theorist Jeremy Bentham theorized a scary idea: the Panopticon. It wasn't an arena in which Chuck Norris could kick-box his opponents into submission. Instead, the Panopticon was an architectural design for a prison, hospital, or other institution that required the constant surveillance of inmates or residents. The structure featured a circular guardhouse at its center with individual cells stationed around it. The cells were walled off from each other but open and completely visible to the central tower, thereby allowing guards to have full, unobstructed views of all inhabitants at all times.

Panopticon-inspired prisons have been built the world over. The best and most faithful example is the Prison Modelo on Cuba's Isla de la Juventud. Bentham's original description of its potential effects sounds pretty creepy: "Morals reformed, health preserved, industry invigorated, instruction diffused, public burthens lightened, economy seated, as it were, upon a rock—the Gordian knot of the poor-law not cut, but untied—all by a simple idea in architecture!"[13]

Panopticism as a concept extends well beyond architecture and has existed for centuries, well before Bentham coined the phrase, in the form of religion. For millennia, religion has affected people's behavior through the specter of an ominous god watching and judging from on high. Fearful millions of people have gone against their base instincts to be, well, nicer. Organized religions have caused their share of historical atrocities, but it does make you wonder whether we might have committed even more unspeakable acts if we didn't have that constraining influence. Panopticism sounds ominous, but there's also clearly an upside to it, as Bentham suggested.

That's the fundamental principle that lies behind the modern world's conversion from religious and architectural panopticism to a technological one. Google and Facebook may have large windows into people's lives, but they're also providing them with fantastic services (well, Google is). Governments may be gathering scads of data on their citizens—against the law in many cases—but it's ostensibly doing so to keep them safe from evildoers. Given what we know about how surveillance affects behavior, the downside is pretty clear: The technological panopticon is forcing us to be good little rule followers. It's a dystopia of uniformity scarier to individualists than any robot war or bioengineered virus.

Hopefully, like the impersonal disconnections enabled and accelerated by technology that we saw in the previous chapter, this is temporary. We're still drunk with what all this new Internet capability and ubiquitous computing is giving us, but we're slowly waking up to several hangovers of what it's costing us. Ian Kerr, the privacy expert at the University of Ottawa, likens the near future to the obesity problem mentioned in chapter three. The Big Data voluntarily generated by people—ultimately accessed and used without their knowledge and consent—is like Big Sugar. We're hurtling toward diabetes. "You think of the person who could have completely avoided all of the health issues they've had if they had just paid a little more attention to the sugar," Kerr says. "I'm kind of worried that we'll find the same thing with information. People are pressing their salience on different things and later they'll see how the databases can connect the dots."

These last few chapters have shown that humanity is moving strongly toward greater individualism, which means that the value of privacy will increase. Recent surveys and reactions to news of government spying reinforce that notion. At the same time, the reaction against technological panopticism will be equally strong.

One possibility is a move toward disconnection, in which people decide that the value of always being online isn't worth the tradeoff. That outcome seems highly unlikely, however, given that it's tough enough, say, to get a job now if you can't use a computer or the Internet. Imagine how hard it would be if a potential employer Googles you and you show up as a blank slate.

A more likely scenario is that privacy will reassert itself in several different forms, either through the graduated services that Kosa prescribed or the dissemination and mainstreaming of strong encryption. So far, only big enterprises such as banks and government departments have been using such protections on their all-important data. But with Moore's Law once again coming into play, strong encryption is getting cheaper and easier to use, meaning those protections are marching inexorably toward mainstream adoption. As of 2014, smartphone apps such as HotSpot Shield and SurfEasy were offering to privatize your mobile traffic in exchange for a few dollars a month.

The likeliest outcome is a hybridized version of that Windows Vista–like graduated privacy system, where people subconsciously and automatically adjust their communications based on their level of sensitivity both technologically and in the real world. It's already happening in reaction to the continuing deluge of news of privacy violations. The phenomenon of file-sharing is a good example. In its early days, many Internet users freely and impudently swapped music and movie files without worrying about whether anyone knew they were doing it. But as it became apparent that copyright holders were monitoring and sometimes reacting to their actions, many people started to use encryption protocols that masked their file-sharing traffic. But not many of those same users applied encryption to all of their Internet use; they only turned it on when they were doing something they knew they shouldn't be doing—a coincidental nod to Google's Eric Schmidt.

People making conscious choices about what level of security to apply to their actions is common in my line of work. Very rarely do contacts or sources provide hard information over e-mail; they either call or request to meet me in person. Harkening back to the days of Watergate and meetings with Deep Throat in dark parking garages, people know to reserve their most private or sensitive communications for situations that involve the least amount of technology and therefore the smallest chance for outside recording opportunities or privacy invasion. Better technological tools such as encryption inevitably will aid in this respect, but so too will a renaissance of good old-fashioned low-tech answers. Proper privacy management will involve a combination of the two.

So far, the Internet has required us to make huge tradeoffs. We've been reaping the benefits of more capability and reach, such as many more online friends, freedom of expression, and job opportunities, but we haven't realized what we've been losing. Unprecedented panopticism is causing us to shrink inward. The declining number of close friends discussed in the previous chapter seems to be the most alarming result. But ultimately, there isn't any convincing reason that we can't have both. We're only just now realizing that possibility, but we need to figure out how to do it.

# 8

# Belief: Are One-Eyed Cylons Myopic?

*Dilbert: Wow! According to my computer simulation, it should be possible to create new life forms from common household chemicals. Dogbert: This raises some thorny issues. Dilbert: You mean legal, ethical, and religious issues? Dogbert: I was thinking about parking spaces.*
  —SCOTT ADAMS, *DILBERT* COMIC STRIP, MAY 31, 1989

In South Korea, Samsung devices have something called DMB, digital multimedia broadcasting, built into them. They have retractable antennas that, once extended, allow them to receive satellite television signals. That, combined with the big screens the company's devices usually have, explains why Koreans prefer to buy local. Besides actually communicating with people, commuters can watch TV while on the subway. Nearly three-quarters of the phones used in the country—an incredibly high market share anywhere in the world—are Samsung as a result.

The Venerable Noh Yu, however, is no ordinary commuter. He's a Buddhist monk and the deputy director of the office of missionary activities of the Jogye Order, a sect that traces its roots back to the religion's first arrival in Korea more than three hundred years ago. As a monk in the most wired society in the world, he's in a prime position to opine about the intersection of technology and religion. But I'm more fascinated by the iPhone he pulled from a pocket hidden in his gray and orange robe.

He explains that he deeply considered his decision to carry the TV-antenna-less device. "Apple is a company where the focus is on apps," he tells me through an interpreter. "Samsung is a company where the focus is on the model. It's a difference of emphasis between the two companies. Samsung has a series of models that come out very frequently and the focus is on getting people to upgrade to the next model each time. But a smartphone is a device where you don't have to upgrade to the next model. You can just upgrade to the next app. The point is upgrading your app, not upgrading your phone."[1]

What he says makes sense. As the category leader, Apple tends to get the best applications designed for its products first. Samsung phones, which use Google's rival Android operating system, generally get those same apps later—if at all. "Koreans see it differently," Noh Yu continues. "They think, in general, that to be a smart user of smartphones, you have to get the new model. But that's not the point at all. The point is having a model and being a smart user of the apps on the model. It's not about changing the phone."

His views and his preference shouldn't be surprising. Apple co-founder Steve Jobs, the man who shepherded the iPhone into existence, was Buddhist, after all. His spiritual sensibilities permeated his approach to products, which is one reason the company was so successful under his watch. He made easy-to-use, harmonious technology of which the ubiquitous computing types, whom we met at the beginning of this book, dream. Noh Yu however, apologizes mildly for his choice of phone. He's no less a patriot than the next South Korean, and he generally likes Samsung, but Apple's phone serves his needs better.

If anyone knows about the wisdom of phones and technology in general, it's him. The Jogye Order—dually housed in a grand, colorful temple in the heart of Seoul for spiritual matters and a small office building next to it for administrative purposes—designs smartphone apps to spread Buddhist teachings and raise awareness of events and activities. They outsource the technical design of the apps, but the Order itself produces the content.

South Korea is the most wired (and wireless) country in the world, with some of the most advanced telecommunications networks going. It's a global power in robotics, and the country's two biggest companies, Samsung and LG, are also worldwide electronics leaders, from televisions to tablets. If somewhere in space aliens are studying humanity's technological prowess to decide when to make contact, they surely have their eye on South Korea, where even the monks are tech savvy.

Noh Yu says that, although technology isn't in his religion's nature, the Order must nevertheless adapt to such advancement or risk being left behind, which is why it decided to create apps. "Buddhism is a religion that has always tried to accept and embrace the particular characteristics of the time in which people are living," he says between sips of tea. "There has been a change from the age of the Internet to the age of the smartphone. Buddhism perceives a need to adapt to that change and to accept those cultural developments, so the development of the apps is part of an effort to adapt to the times and make it more accessible and relevant to people today."

But is technology affecting the very essence of religion and spirituality and in the larger sense what people believe about themselves and the world? Websites and apps that provide information and act as recruitment tools are one thing, but is technology helping with the core purposes and values of religion or Buddhism specifically in this case?

"Technology will not have any effect on the Jogye Order or Buddhism in general in the sense that Buddhism is not seeking to change the outside world, but rather the world of the heart," he explains firmly. "There are ways in which technology affects us emotionally and those ways are usually negative. Technology makes people lonelier, they become more distant from each other."

His assessment cuts to the heart of the past few chapters. Technology is making the world a better place in terms of rising prosperity, better living conditions, increasing opportunities, and declining conflict, but its deeper effects on us as individuals aren't as obviously benign. As we saw in the previous two chapters, Noh

Yu may be right when he suggests that technology is making us withdraw from one another, seeking ever more intimate relationships and deeper levels of privacy.

In that sense, it's hard to say we're coming to know ourselves better and, therefore, becoming better people as a result. By most definitions, "better" would cover improvements in such fundamentally "good" characteristics as love, friendship, loyalty, selflessness, caring, and compassion. So then, if all this technological progress isn't leading to improvements in those areas, what's the point of it all? Is a world where everyone lives for two hundred years and is gainfully employed all that desirable if we're all becoming miserable, selfish, soulless curmudgeons?

It's difficult if not impossible to measure these intangible qualities, so it's equally hard to come to any definitive conclusions. But let's take a look and see what we can discover.

## GOD BLESS THE USA

Religion is a good place to start since all of the world's big, spiritual organizations generally encourage adherents to know themselves as well as some over-arching god, and to practice overall "goodness." It's also relatively easy to measure religion itself.

Religious belief began in the time of Blarg as a simple means for explaining the inexplicable, which was at the time pretty much everything. Those crippling intestinal pains that kept our intrepid caveman home and away from the grind of hunting and gathering? Obviously the work of some evil, unseen entity. The lightning that crashed down onto the tree and set it on fire? Clearly the result of some sandal-wearing dude on a cloud tossing electrical bolts at the Earth. Parking on driveways and driving on parkways? Must be the work of the Devil.

As humanity evolved and societies formed, thinking became more complex and reason took hold. As priestly castes collected and organized beliefs, religion morphed into the embodiment of shared cultural values, a central repository both for inexplicable and supernatural events and for rules that promoted societal order and betterment. In many cases, religion inspired law and vice versa. The ten commandments, for example, tell us it was neither socially logical nor acceptable to steal from or kill one's neighbor. For a great stretch of recorded human history, religion and law indivisibly intertwined, from the Roman Empire—where the emperor also headed the state religion—through to the divine right of kings. That's still the case in some parts of the world today. Religion now varies in its nature and, depending on the country, spans this spectrum of development. In the most primitive societies, religion still acts as the primary explanatory force of natural occurrences. In theocracies, it's the governing body. In many advanced countries, it's less a congregation of the faithful than a community group.

But we shouldn't underestimate religion's power as a cohesive force. My family, for example, is Polish, and without religion—the Catholic Church specifically—the concept of "Poland" may have ceased to exist long ago. Carved up by neighboring powers and wiped entirely from the map three times, Poland—along with its customs, language, and traditions—might have been assimilated by its conquerors over the centuries and ultimately faded away had the Church not preserved the idea of it. When Poland coalesced again geographically and politically, the nation was still culturally intact. That does much to explain why the country is so predominantly Catholic today—more than 85 percent of the populace—and why, even though I'm not religious, I can see the positives of such an organization.

Religion's negative side tends to come to the fore in places where it has been strictly institutionalized, which includes Europe through much of its history as well as large parts of the current Middle East. In such cases, it was or is a tool for explaining, maintaining, and enforcing social orders. Why are people poor and suffering? Because God wills it! In its later incarnations, religion often justifies or disguises the perfectly explainable failings of human nature. That goes double for the untold death and deep suffering it has caused. Dangling the carrot of the afterlife to the masses offered an exchange for their obedience and misery. Sure, you might have to eat dirt now, but after you die you'll be drinking mead from golden chalices and golfing all day, every day.

Nevertheless, as prosperity has grown over the past few centuries, religion has seen a marked decline in advanced countries. As just a few examples, over the second half of the twentieth century, belief in God declined by a third in Sweden, a fifth in Australia, and about 7 percent in Canada.[2] Researchers have gone beyond calling this a correlation to declaring it a full-on causality, where the growth of prosperity definitively leads to a decline in religious belief.

The numbers back up such claims. Close to half of people living in agrarian societies say they attend church at least once a week compared to just a fifth in advanced societies. In nations where per capita income falls below two thousand dollars a year, about 95 percent of the population say that religion plays an important role in their lives. Bangladesh leads the pack, with a 99 percent affirmative response rate—and life is probably pretty tough for that outlying heathen 1 percent.[3] In countries where per capita income falls above twenty-five thousand dollars or more, the percentage is half that.

By this measure, Estonia is the least religious country in the world. Only 16 percent of people there say religion is important. But Estonia's championship in this category is not entirely fair since the Soviet Union banned religion there for many decades. Adjusted for totalitarianism, the crown really belongs to the kingdom of Sweden, which counts only 17 percent of its populace as religious despite having no restrictions on beliefs. On the flip side, according to WIN-Gallup's 2012 International Religiosity and Atheism Index, the top ten most religious countries—where people self-identified as such—all had per capita income of less than fourteen thousand dollars, with Ghana, Nigeria, and Armenia leading the pack.

"It's only popular in societies that . . . have enough rate of dysfunction that people are anxious about their daily lives, so they're looking to the gods for help in their daily lives," says paleontologist Gregory Paul. Religion thus directly measures a society's stability and prevails most in places where life is tough in terms of conflict, health, and poverty. "It's not fear of death that drives people to be religious, and it's not a God gene or a God module in the brain or some sort of connection with the gods," says Paul. "It's basically a psychological coping mechanism."[4]

In poorer countries, religious institutions tend to provide essential services such as education, as well as social networks that help people cope with trauma. "Religion becomes less central as people's lives become less vulnerable to the constant threat of death, disease and misfortune," write sociologist Pippa Norris and political scientist Ronald Inglehart in *Sacred and Secular: Religion and Politics Worldwide*. "As lives gradually become more comfortable and secure, people in more affluent societies usually grow increasingly indifferent to religious values, more skeptical of supernatural beliefs and less willing to become actively engaged in religious institutions."

But two outliers buck this rule: China and the United States. China, although advancing quickly, still has relatively low per capita income—less than eight thousand dollars per year—yet it also rates very low in many religiosity measures. It's technically the least religious country in WIN-Gallup polls, with only 14 percent of the population identifying as such. But we know why: The Chinese government, at the height of its totalitarianism phase in the 1960s and 1970s, effectively outlawed religion and destroyed many places of worship. Only in recent years has the Communist party there relaxed restrictions, allowing for traditionally Chinese religions such as Buddhism to make a resurgence. But increasing prosperity is countering religious growth. In fact, China's leaders have criticized their own society for being too secular. Talk about irony.

The United States is a more interesting case because it's one of the world's wealthiest nations, yet it also ranks high in religiosity. More than 80 percent of Americans say they believe in God and always have, a shockingly high number compared to, say, France at 29 percent or Britain at 36 percent. Almost a quarter of Americans say they attend religious service once a week, compared to just 5 percent in France and 10 percent in Britain.[5] Proof of the phenomenon also lies in the election speeches of just about any American politician: God is almost always invoked.

Sociologists say that, although the results look surprising, they do add up and correlate. Religion flourishes in places of turmoil, uncertainty, and especially poverty. The most religious of the states in the union—Mississippi, Alabama, and South Carolina—are also among the poorest. The least religious—Vermont, New Hampshire, and Maine—are among the wealthiest. The United States also has the highest level of inequality among developed nations. As we'll see in the next chapter, this metric deeply affects general happiness and hopes for the future. "We have fifty to sixty million people without health insurance; we have the highest child poverty rates of the industrialized democratic world; the greatest gap between rich and poor of the industrialized democratic world; we have increasing inequality and, voilà, we also have a strongly religious society . . . That can't be accidental," says Pitzer College sociology professor Phil Zuckerman.[6]

Religion will continue to decline in proportion to the growth of prosperity and equality. Northwestern University engineering and math professors Daniel Abrams and Haley Yaple have created a model projecting the decline of religion. After studying competition for membership between various social groups in eighty-five religious and nonreligious societies, they concluded that extinction is inevitable. "The model indicates that in these societies the perceived utility of religious non-affiliation is greater than that of adhering to a religion, and therefore predicts continued growth of non-affiliation, tending toward the disappearance of religion."[7]

# THE PASSION OF THE YODA

People shy away from religion in developed countries for additional reasons. Among them is displeasure with doctrines that don't change for centuries. Unlike the Jogye Order, some large religions aren't willing to adapt to changing times and surroundings, frustrating many would-be adherents to the point of exit.

Many people are surprised to learn that Anne Rice, our vampire novelist from chapter three, is quite religious. Aside from her best-known work, she has written several books about the adventures of Jesus Christ. But she also counts herself among disillusioned believers who have no kind words to say about organized religion. "I have a great faith in God and great personal devotion to Jesus Christ, but not to the Christian belief system, not to any theological system made up years after Jesus left the Earth that argues all kinds of things about which he said nothing," she tells me. "Many people feel that way today. They have their devotion to Jesus and their faith in God, but they are way past listening to any church rant to them about how they shouldn't use artificial birth control or they shouldn't go to their gay brother's same-sex wedding. We're through with that nonsense, a lot of us. We're just not going to take that kind of superstitious dictation."[8]

For the purposes of our analysis of religion and spirituality, she says it doesn't have to be a zero-sum game. Individuals can be both secular and spiritual without having to belong to an organization. It's possible to subscribe to some of the loftier goals prescribed by religions without having to adhere to their more questionable positions. "I wouldn't equate secularism with the loss of spirituality, not by any means," she says. "The secular people I know have been deeply spiritual and deeply concerned with morality. They have always had conscience and deep profound values. Many of us, in getting away from religion, have defined ourselves at times as secular, but that does not mean we don't have deep values and a deep spiritual life. Churches don't own spirituality."

Shane Schick, a friend and fellow journalist, worked with me at *Computer Dealer News*. I left for jobs at various newspapers while he stayed and ultimately took over the place. I learned only in recent years via Facebook that he's always been religious. We met for coffee so I could ask how he reconciled those two aspects of himself, his professional side—which deals with science, technology, and the advance of reason—and his spiritual side, which believes in the unprovable. His answers were surprising.

Like me, Shane was born into a Roman Catholic family, and he went to a Catholic school. But for him, church and religious life were about providing service to the community. He recalls proudly how his grandparents rebuilt the doors on their local church. It wasn't until he went to college that he realized he lived in a secular world. He fell out of religion somewhat during those years, until he met his wife-to-be (online). At the time, she was earning a master's degree in theology. "I don't know how other men would react, but that attracted me immediately," he says with a laugh.

They hit it off, but they encountered a potential roadblock: She was Anglican. It didn't turn out to be an issue in the end since he found that their two Christian sects were actually hewed quite closely to each other. He came to prefer Anglicanism, so he switched. "Anglicans treat themselves a little more like a work in progress," he says. "There's a lot of questioning. I enjoy that. I find it accommodates doubt a little better."[9]

That, in a nutshell, is how he reconciles being a devout Christian and a technology journalist at the same time. The first makes him better at the second. He explains that he specializes in advising his readers—generally business executives looking to make decisions about which technologies to pursue for investment—on good and bad bets. To do that, he needs to cast doubt on the endless stream of products that vendors foist on him. "I think you need a healthy skepticism," he says. "If I was someone who immediately got obsessed with every new gadget, I think that would be a disservice. You can't be strategic if you just uniformly go with whatever hype is coming out of the mouths of consultants. That dubiousness is important."

Technology journalism is rife with fanboy-ism, so never mind Jesus—having doubt as your co-pilot is a good idea in this line of work.

When it comes to religion, though, it seems contradictory. "Aren't doubt and faith opposites?" I ask. Not necessarily, Shane replies. "The enemy of faith is not really doubt, it's more indifference or apathy." Religion for him isn't about blind devotion but about seeking answers. You can only do that if you ask more questions.

Our conversation was verging into Yoda-isms, which probably isn't surprising given the subject matter. It's also a little amazing that Shane is able to have such a lucid conversation. We're sitting in a Starbucks with his infant daughter literally crawling all over him. I ask how he can manage articulating coherent thoughts while this is going on. "This is nothing," he replies, Zen-like, and continues.

"I can't prove God exists or half the things that Christians believe in," he says. "At the same time, I do believe technology can bring us closer to some answers. Technology at its core is a pursuit of truth. Faith is similar. They are just interpreted differently and have different expectations placed on them."

He illustrates the point with an apples-to-apples comparison. "If you're in a room with a bunch of people looking at a software report, you can't be sure of everything that's in there. But as a team, you can still make decisions because you take it on faith." A fascinating point.

We veer back to discussing why people are turning away from religion, and here Shane suggests another, largely technologically driven reason: the rapid and huge growth of individualism. The exponential increase of communications and self-expression that we saw in chapter five, the corresponding withdrawal of the self in chapter six, and the ever-increasing desire for privacy in the previous chapter all point toward the same development: People increasingly are identifying themselves as individuals. They are putting themselves "out there" more than ever before, but they are also devoting more energy to building and maintaining their own identities, separate from the hordes. It's the paradox of the Internet age, writ large: In a world growing larger every day, it's becoming increasingly important for people to establish their own small place in it. "Our culture is also one that is pulling away from a sense of community in general," Shane says. "Forget organized religion, what happened to the Kiwanis and the Legion and all those things that everybody used to belong to? They barely exist anymore except for old people. Our culture is far more individualized. We're not really joiners anymore."

Underlining the point is the sea change that's happening with what people are choosing to do with themselves after they die. Cremations are on the rise in many countries for a number of interlinked reasons, starting with the globalization discussed in chapter two. People increasingly are buying cheaper coffins made in China, which is causing profit margin decline among service providers in the West. That's causing the industry itself to contract. In the United States, nearly twenty thousand funeral homes operated in 2011, about 10 percent fewer than a decade earlier.[10] Tied to this trend is our increasing mobility, which gives rise to a number of questions when a death occurs. When someone dies now, is she buried in the town where she was born or where she spent most of her time? What happens if her family wants to move away from the chosen burial location—do they take the body with them or leave it where it is? Even if the urn is buried, cremation solves these dilemmas because it makes people's remains more mobile as well. "People are less rooted in a community than they were a generation ago," says Patrick Lynch, president of Lynch & Sons Funeral Directors in Detroit. "People raised their families in a community then stayed and died there, but baby boomers move from city to city and from job to job. With cremation you have more portability."[11]

Fewer individuals are finding that it makes sense to go through the often exorbitant expense to have a big funeral, even though cremation goes against traditional beliefs in the Jewish, Christian, and Muslim faiths. The Catholic Church, for one, outlawed cremation until 1963 and continues to urge its adherents to stick with burials. Still, in the United States, logical economics triumph over the country's relatively high religiosity in this area. In 1985, only about 15 percent of deaths were followed by cremation, a number that had risen to around 40 percent by 2012.[12] But the US lags considerably behind many other advanced nations, which isn't surprising given its strong religiosity. The countries of Scandinavia, as well as Canada and the United Kingdom all surpass 70 percent, and land-starved Japan cremates almost everyone. Cremation typically costs only a few hundred dollars, as opposed to the thousands required by funerals, so many families and individuals find it's not worth having a lavish ceremony, religion be damned.

## THE PARADOX OF ANSWERS

Buddhism, which emphasizes the annihilation of the self, generally favors cremation over burial, but back in Seoul the Venerable Noh Yu isn't so sure whether growing individualism is a good thing or if it's making people better. He became a monk to escape the secular world and the "cycle of life and the pain and suffering that people experience as they live." Only by abandoning the technologically driven, secularly prosperous world did he himself become a better person. Technology makes our lives more convenient, but then we become trapped within that convenience and lose sight of what really matters. He describes the process of taking a photograph. As discussed in chapter five, it used to require significant effort, but, now that we can take instantaneous pictures with our phones any time we want, they're not as meaningful. We store thousands of them on computer servers and never look at them, whereas in the past we might pay twenty dollars to print one, frame it, and hang it on the wall, where we'd look at it every day and potentially contemplate it or remember how we felt when it was taken.

In that way, technology can't help us come to know ourselves because it distracts us from that deep contemplation, he says. Aside from mood-altering drugs, technology cannot change or control or fix our feelings. "When you were coming here today, you could decide, 'Are you going to come or not?,'" he says. "Technology can help you get here, but it can't help you decide whether or not you're going to come."

He clearly has given this topic a lot of thought, but that's what people who meditate do. "You're also losing yourself because it's through the people that we're close to that we're able to understand ourselves. How they react to us, how they see us, what they say to us—you can see that as a mirror of the self. That's something that's forfeited as we head in that direction," he says.

"When you deal with people around us and close to us, you have to accept them for who they are. There are going to be differences, incompatibilities, but those are things we have to overcome. It's a crucial part of human relationships and it's something that doesn't happen in relationships that develop only online. You can't live with people online. It lacks the true nature of acceptance."

In one of those incompatibilities, he and I have a difference of opinion in how the seeking of answers happens. If science, and therefore technology, is advancing exponentially, won't it eventually answer all of the universe's questions? When that happens, won't the mathematical prediction of religion's extinction come to pass?

Both Noh Yu and my friend Shane had the same response. Science is indeed answering questions at an accelerating pace, but it inevitably produces more questions. Noh Yu brings up the placebo effect. Medicine can solve certain health problems, but individuals seemingly heal themselves of certain ailments if they believe they're taking something that will make them better. "What's the scientific principle behind that?" he asks. "The nature of the soul is very complicated, and in the end it can never be wholly explained or analyzed." Shane mentions other historic discoveries, such as when we figured out the Earth wasn't flat, or when we split the atom. "We thought that we had made the ultimate discovery, yet there was more," he says. "There are some philosophical questions that were debated thousands of years ago that are still being debated today. They're just not easily answered and maybe can't be answered."

It seems that questions, like answers, also multiply according to Moore's Law.

## THE UNIVERSAL ALARM CLOCK

Futurist inventor Ray Kurzweil, whom we met in chapter one, is one of the biggest adherents to science on the planet. He believes everything in the universe—including people and our souls—is a pattern waiting to be discovered and explained. Once that happens, all answers will be laid bare. As neuroscience advances and the brain is understood more fully, he's confident that we'll learn everything about what makes us human: not just what makes us tick, but also our personalities and the very essence of our beings. Once a pattern is understood, it can be replicated and then duplicated. Theoretically, this can happen in other "substrates," his fancy word for bodies or platforms.

In plainer English, once we figure out the patterns that make up our personalities—and therefore our core being or soul, as it were—we'll be able to copy them and upload them into our choice of sources: a biologically cloned body, a mechanical robot, a virtual world on a computer server, or some hybrid of all of the above. In any case, we'll be able to live forever.

Singulartarians such as Kurzweil also believe that as computers surpass human intelligence—around the year 2045 according to their interpretation of Moore's Law—it's only a matter of time before they expand to become the essence of the universe itself. By this estimation, we're in the third-last epoch of evolution, where technology is prevalent in the form of hardware and software. The next epoch, which will begin once we cross the Singularity threshold, will be the merging of human intelligence with technology. If you can't beat 'em, you have to join 'em. The benefits of slowly but surely merging our brains with technology will become increasingly apparent. If you think that suggestion's nuts, try avoiding Google for a month. It will become clear just how many of our brain functions we've off-loaded to technology already. Joining more fully with technology on a biological level will also help keep that same technology from wiping us out in a *Terminator*-style Armageddon.

In the final epoch, the universe "wakes up" as all matter and energy are infused with hybrid human-machine intelligence. What happens then? That's when things get interesting and loop back to the question of God and the answers of the universe. "One thing we may do is to engineer new universes," Kurzweil says. "Similarly, our universe may be the creation of some super-intelligences in another universe. In that case, there was an intelligent designer of our universe—that designer would be the evolved intelligence of some other universe that created ours."[13]

On a less meta-cosmic scale, we're getting closer to answering the question of just who we are, and it's happening outside religion and religious belief—maybe even in spite of it. The evidence, gleaned over the past four chapters, is convincing. Communication between people and their own individual expression, whether through status updates and photos or self-made music, videos, and games, is exploding. So is the scale on which we socialize technologically. At the same time, we're also withdrawing into ourselves, valuing our privacy and intimate connections more. We're also pulling away from long-held institutions such as marriage and religion. These aren't just superficial developments; they indicate that we've entered a new Golden Age of individualism. You might consider that a bad development—by which we're becoming the sort of miserable, selfish, soulless curmudgeons I invoked at the beginning of this chapter—but as the next chapter illustrates, this is in fact a good thing.

# 9

# Happiness: It's Always Sunny in Costa Rica

*Money can't buy you happiness, but it can buy you a yacht big enough to pull up right alongside it.*
— DAVID LEE ROTH

Big bucks, long life, realizing who we are—technology must be making us amazingly happy, right? Well, that depends on what we count as happiness.

Happiness consists of at least two different concepts. One is a longer-term feeling or sense of general satisfaction. It can be conscious or subconscious, and it can change depending on whether you're thinking about it. You might go contentedly about your daily routine and not really notice it, but when you sit down and consider your life, you might realize you really aren't happy with it. It can also work the other way around. Maybe you're usually a miserable cur, but when you stop to take stock you realize that life is actually pretty good. Either way, satisfaction with life is an ephemeral quality that we all contemplate from time to time in different ways.

The other flavor of happiness is more short term. It comes in bursts and is generally more extreme than the longer-term type. It happens when someone tells a funny joke, when you plunge down that incline on a roller coaster, when you orgasm, or when you trip out on drugs. The elation feels strong, but it often fades after a time. You may remember it for a few days afterward, but otherwise it becomes just another fond memory.

The age-old nature-versus-nurture dichotomy ultimately governs and influences both types of happiness. Studies have found that genetics and brain chemistry determine up to half of our potential emotional happiness, whether fleeting or long term. We *can* get happier in this regard with psychological treatment or with external stimulants such as antidepressants or narcotics.

This particular kind of happiness will continue to improve as our understanding of the brain deepens. With new discoveries, we'll have more ways—drugs, psychological treatments, and possibly even mental tricks—to make ourselves happy. Then, of course, the question will come down to whether we'll actually bother with them or not.

Neuroscience is one of the fastest accelerating sciences. Its practitioners are routinely challenging long-held theories. For much of the twentieth century, for example, the consensus held that after a period of flexibility in childhood the brain was relatively immutable. Science thus supported the theory that people become more set in their ways the older they get and that you can't teach an old dog new tricks. Recent research, however, has cast considerable doubt on this line of thinking. Many scientists now believe in neuroplasticity, the ability to change both the brain's physical structure and functional organization.

It's not as complicated as it sounds. Simple tricks are often able to change behaviors. With just a box and mirrors, neuroscientists at the University of California at San Diego have succeeded in relieving amputees' "phantom pain," a sensation that manifests as itching or intense discomfort in places where they used to have limbs or appendages. The box has two compartments, one for each arm or leg. Placing the mirror in the middle of the box creates a reflective illusion for the amputee of seeing a whole limb in place of the missing one. Through simple movement exercises involving the whole limb, patients subconsciously fool their brains into thinking that they still have the absent part. Painkillers and surgical treatment had little success in treating such phantom pains, but the box trick has worked remarkably well. In 2007, the US Army tried it on a group of amputees, and all of them reported a reduction in pain. The project was then extended to Vietnam, Cambodia, and Rwanda to help treat land mine victims and people afflicted with leprosy. Similar results followed.

Neuroplasticity will play an increasingly important role in treating all manner of addictions, from drugs to gambling. Scientists have surmised that addictions happen when certain neural pathways become inflexible through prolonged indulgence. Imagine the neural pathway as an electrical cord and a particular drug as an outlet. Addiction happens when you can't pull the plug from the wall, thereby making the connection itself rigid and immutable. Pain, depression, anger, and a loss of self-control result unless the habit is fueled continually. External treatments such as quitting cold turkey or even other drugs may help in the short term, but they often don't alleviate the long-term issue because the neural pathways exist as before, which can result in relapse. "We are probably not going to find new therapies by trying to understand the modifications caused by a drug in the brains of drug addicts since their brain is anaplastic," wrote researchers Pier Vincenzo Piazza and Olivier Manzoni, from the Neurocentre Magendie in Bordeaux, in a 2010 report. "The results of this work show that it is in the brain of the non-addicted users that we will probably find the key to a true addiction therapy . . . Understanding the biological mechanisms which enable adaptation to the drug and which help the user to maintain a controlled consumption could provide us with the tools to combat the anaplastic state that leads to addiction."[1]

The Brain Activity Map launched by the US government in 2013 is expected to net the same kind of breakthroughs in neuroscience as the Human Genome Project did and will continue to do for plain old biology. Beyond that, there's also neural engineering. It won't be long before scientists can identify and then eliminate or lessen the impact of those parts of us that make us sad or unhappy.

The question then becomes—as it does with all scientific methods of instilling happiness—whether we *should* use them.

As with all potentially transformative technologies, there is considerable consternation about neural engineering. Mirroring the fears of genetic manipulation and designer babies—where parents pick and choose which traits, such as gender or eye color, they want in their kids—critics worry that neural engineering will lead to the wholesale elimination of certain behaviors and thought patterns that we don't understand fully. Depression and bipolarity, for example, could be eliminated, even though recent studies have found that people who suffer from these conditions are often more creative.

Such fears are overblown, though, because they don't acknowledge how such technologies are absorbed into society. They don't arrive suddenly, but rather incrementally over a long period of time.

Robots are a good example. Society didn't go from computers to cyborgs in a single year or even a century. Robotic technology has been creeping into our lives slowly. Take the robot cars predicted to hit the mainstream by 2020. They won't just appear from the clear blue sky then, fully autonomous and ready to go. Each year's new models add one more piece of that eventual puzzle. Carmakers have been doing it for decades by slowly adding cruise control, lane detectors, rear cameras, and even automatic parallel parking abilities. This incremental transformation allows plenty of time to assess, discuss, and prepare for new paradigms. With the world barreling toward fully autonomous vehicles, the conversations about what changes they'll bring are already happening now. (It's also the main reason that Singularity University exists.)

The initial applications of neural-engineering will be, like bioengineering, therapeutic in nature. Once scientists have a fuller understanding of how to change the brain, demand will call—with virtually no resistance—for applying such techniques to debilitating conditions such as Alzheimer's, Parkinson's, and cerebral palsy.

During this phase, we will have plenty of opportunity to debate the scope of the technology and how far it should go. Should it treat or eliminate depression? What if people want to use it recreationally?

But technology can sometimes move too quickly for such conversations to take place, which squares with what we know about Moore's Law. The NSA's wide-scale privacy violations, discussed earlier, stand as good examples of technological capabilities running amok and forcing society to play catch-up with its values and laws. Technology automatically outpacing social reaction to it may be the new status quo, which certainly is cause for concern because it necessitates a new mode of collective thinking. Transformative technologies need to be assessed early in their development processes. We have to start thinking about the ramifications sooner and faster. Fortunately, our ability to gather data, have conversations and come to evidence-based conclusions is also improving quickly, thanks yet again to Moore's Law. So it may be just the legal aspect that needs to speed up.

But lest we forget, only about half of our happiness is internalized. The rest depends on the world around us and what we do to it. Biological happiness will improve, so are we getting happier according to nonbiological measures? In a word: Yes. As happiness experts put it—and, yes, happiness experts exist—health and life expectancy are the biggest affective factors on positive happiness levels. No matter what fatalists or goths may say, people aren't happy when they're sick or dead or when everyone around them is. In that sense, happiness is increasing in lock step with prosperity and life expectancy. As the United Nations' 2012 World Happiness Report puts it:

> *Life on earth has, at least on average, become much less brutish, nasty, and short over the past five hundred years. The evidence for this ranges from falling murder rates to rising life expectancies. There are no long-standing happiness measures available to track these life improvements, but it would*

*be no surprise if individual and community-level aspirations
and standards have risen over the same centuries, even if at a
lower rate.*[2]

Just as with health, happiness ties to economic prosperity. Where one increases, so does the other. However, happiness shares further similarities with economic growth and, in particular, what's happening with global equality. The world is becoming happier, but happiness within certain countries isn't necessarily keeping pace. Some countries, in fact, have stalled. The United States, despite amazing economic growth over the past half century, has seen little to no improvement in reported life satisfaction levels over that time.

Rigidly thinking economists believe that happiness is linked solely to financial prosperity, but they're wrong. If people were purely rational, then money *could* buy happiness . . . and Justin Bieber and *Jersey Shore* would never have become cultural icons.

Many other factors besides income come into play. How secure people feel, whether they have other people around whom they love and trust, how long their commute is all determine an individual's happiness. Maslow's hierarchy of needs states that physiological necessities—breathing, food, water, and sleep—plus safety issues such as security of resources, employment, and health must be fulfilled before anyone can approach life satisfaction. More ephemeral factors build upon that baseline, including love, belonging, friendship, intimacy, confidence, achievement, having the respect of others, creativity, spontaneity, and self-actualization. All of those are necessary for humans to achieve true happiness and life satisfaction. As we've seen, technology plays a role in every one of those areas.

## MAKING MORE THAN THE JONESES

A number of ongoing efforts exist to measure happiness, including the Gallup World Poll and the European Values Survey.

Happiness research, unfortunately, suffers from the "giggle factor." There's a tendency not to take it seriously. But we should pay attention to such studies because researchers have been refining their methodologies for decades, formulating and testing more incisive questions every year. Such social scientists are benefiting from their own version of Moore's Law, where the efficacy and accuracy of their surveys are improving. This isn't just touchy-feely nonsense.

According to the Gallup World Poll's Cantril ladder, which measures overall happiness levels from one to ten (ten being the highest), about half the world rates moderately happy, scoring between four and six. Fewer than 3 percent of people rate themselves as very unhappy, a one, while almost a similar proportion is extremely happy, at ten. The United States, Canada, Australia, and New Zealand have the highest percentage at eight or above—almost a third of their respective populations—while at the other extreme lies sub-Saharan Africa, where almost a third of people are at three or below. The top five happiest countries in descending order: Denmark, Finland, Norway, the Netherlands, and Canada. The United States still ranks relatively well at tenth, despite its lack of improvement in recent years.

Given the highest-scoring countries, the correlations with economic prosperity are obvious. But then what? Income obviously plays a significant role, but the Easterlin Paradox—coined by economist Richard Easterlin in 1974—eventually comes into play. His theory holds that richer people are generally happier than poorer people, but they don't get much happier as they get even richer. This is also known as the diminishing marginal utility of income. An extra dollar improves a poor person's satisfaction much more than for someone who is already rich. Put another way, a ten-thousand-dollar raise for someone making ten thousand dollars a year results in a measurable increase in happiness, yet a ten-million-dollar raise to someone already making ten million probably won't make much of a difference to their overall satisfaction—apart from a brief, euphoric spike that might involve a big party on a yacht.

The hedonic treadmill theory holds that people generally return to a relative state of happiness after major events or wind-falls. So that extra ten thousand dollars makes somebody happy temporarily, but if he was a miserable sort to start, eventually, even as a twenty-thousand-dollar earner, he'll return to being just as crusty as before. In this case, happiness is a social measure that's relative compared to how everyone else is doing. Ask yourself this question: Would you rather make a hundred thousand dollars a year if that's what everyone else was making, or would you prefer seventy-five thousand if everyone else was getting only fifty thousand? Many people would take the lower amount because it means they're surpassing the Joneses.

These theories explain why some poorer countries do well in the various happiness studies. Costa Rica tops average life satisfaction in Gallup Polls, while Puerto Rico leads the same measure in the World Values Survey. Bhutan, a tiny kingdom nestled between Nepal and India, has an average life satisfaction of six out of ten, despite having a per-capita income of fewer than three thousand dollars. The rest of the explanation falls to those other factors from Maslow's hierarchy of needs—the fuzzy stuff like friends, family, love, and cute bunnies.

Social capital—such as the relationships and trust discussed in chapters six and seven—is very important to happiness, and it's been tested in various ways, most notably by *Reader's Digest Europe* in 1996. The magazine performed an experiment in which it dropped ten cash-bearing wallets, which included identifications with names and addresses, in twenty different Western European and a dozen US cities. The frequency with which the wallets were returned, cash intact, correlated highly with previously estab-lished levels of national trust in international surveys. In Canada, meanwhile, surveys have found large measures of trust for neigh-bors, police, strangers, and co-workers, which helps explain why

Canadians rank well in happiness ratings, and why they value privacy so much. We trust each other not to violate it, so an erosion of privacy equals an erosion of trust.

Trust has fallen substantially over time in countries such as the United States and United Kingdom and has risen in places such as Denmark and Italy. The United States hasn't seen its overall happiness levels increase in decades. By way of explanation, the World Happiness Report points to "a decline in the quality of human relationships, as measured by increased solitude, communication difficulties, fear, distrust, family infidelity and reduced social engagement." Basically, everything we've discussed in the past few chapters.[3]

Equality is also an important measure, and, as we've seen, it's worsening in some places such as, again, the United States. Greater inequality leads to increased social tensions—Occupy Wall Street, for example—more poverty, and higher crime. Those problems build upon themselves, with racial unrest, endemic corruption, and police abuses following, all of which contributes to an erosion of social trust.

Ironically, our deepening individuality may be having a detrimental effect on happiness. Take marriage, for example. People already happy on their own have a higher chance of attracting a partner, and statistics show that marriage boosts happiness, at least for a time, because such unions typically fulfill many of Maslow's needs: a spouse provides love, support, a buffer against life's shocks, sex and intimacy, plus a financial partner. Studies also have shown that substance abuse occurs less often with married people.[4]

Conversely, divorced people tend to be happier than unhappily married people and having children—perhaps surprisingly—has no net benefit on life satisfaction levels. Kids may bring joy to a person's life, but they also entail economic commitments and a significant loss of personal time. That changes as children get older, but as my own mother cheekily says: That's when the disappointments start. Your little bundle of joy may not take up your time or money when he's older, but he could end up dealing drugs, joining a gang, or not calling every couple of days.

Some evidence suggests that declining religious belief may be affecting happiness levels too, although this is something of a paradox. While there generally is, as the World Happiness Report puts it, "strikingly more positive emotion and less negative emotion among those people who are more religious" in countries where life is tough, prosperity counters some of that. In other words, having your own home and a reliable source of food beats believing in God any day. Even still, greater religiosity has some correlation with fewer depressive symptoms, and most studies find at least some positive effect of religion on well-being. Which makes sense. Most religions warn against materialism, another word for the hedonic treadmill that opposes happiness.

Religion also helps breed altruism. Most faiths encourage adherents to help their neighbors. Plenty of evidence proves that people who are more selfless are happier. In fact, the links go beyond correlation to causality. For example, many volunteering opportunities arose after the fall of the Berlin Wall in 1989. When the two Germanies formally reunified a year later, those opportunities ended abruptly. People who had volunteered in the East experienced bigger drops in happiness levels than those who hadn't. Another experiment had a group of test subjects undertake three extra acts of kindness to strangers per day, and found that significant boosts in happiness resulted.

The bad news, according to the World Giving Report, is that charity—whether donating money or time or simply helping a stranger—has declined. The study measures incidents of charity, not currency amounts, and covers 146 countries. In 2011, "hundreds of millions fewer people have helped others than was the case last year. This has inevitably resulted in a dramatic reduction in charitable support for millions of vulnerable people the world over."[5] Between 2007, the first year of the survey, and 2011, donations of money, volunteering, and helping strangers each dropped several percentage points. The report authors couldn't explain why this had happened, other than speculating that the global economic crisis may have curtailed people's generosity.

The most generous country in the world by this measure is Australia. Two-thirds of the population down under donate money and help strangers, while a third volunteer their time. The top ten are a relatively diverse bunch: Ireland, Canada, New Zealand, the United States, the Netherlands, Indonesia, United Kingdom, Paraguay, and Denmark. Over the survey's five years, countries such as Liberia and Sri Lanka have also ranked near the top.

The good news is the report counter-proves the more recent trend because charity appears strongly linked to economic prosperity. A few developing countries rank among the most giving, but prosperous ones dominate the list. We can assume, then, that a country generally becomes more giving as it gets richer, despite periodic fluctuations. As the world becomes more economically prosperous, charity will continue increasing across the board as well.

When we put all these measures together, a clearer picture of the causes of happiness—and the role that technology plays in it—emerges. Economic prosperity and good health provide the baseline upon which more intangible needs can be met. On the flip side, erosion of trust, envy of others, the decline of group institutions, and the rise of individualism act as counter-forces to long-term happiness. The world as a whole is becoming happier as more of it moves toward the baseline, but then what? Is there a point to all of our technologically driven prosperity and good health if they don't move us beyond that baseline?

The importance of that question proves the validity of happiness surveys and the measure of life satisfaction in general. It makes sense for less happy nations to look to happier countries to see how they're doing it. In other words, we should all be looking to Denmark. The tiny Scandinavian kingdom of just six million people ranks atop most happiness surveys thanks to an impressive array of world-leading statistics in various measures. It's wealthy, ranking sixteenth in gross domestic product per capita, and it's one of the most equal countries in the world, ranking first to fifth in terms of the Gini coefficient depending on who's doing the measuring.[6] Denmark is also first in entrepreneurship and opportunity. Starting a new business is easy there, so people do it a lot.[7] Danes are thirty-third globally in life expectancy, although the seventy-nine years they typically live falls only a few years short of world leader Japan's eighty-three. They don't get married much, having the eighth-lowest rate in the developed world. Denmark is also the third least religious country in the world, with only 5 percent of the population admitting to attending church.[8] So how did the Danes arrive at this curious stew of happiness, and what role does technology play in it?

## VIKING GAME THEORY

After exchanging hugs and hellos, my friends in Copenhagen lead me to the train into town, which is a robot. No human drivers operate it, which results, I'm told, in highly efficient and on-time service. The only downside is that you have to hurry on because the doors close promptly. A system run by efficient yet heartless machines doesn't allow time for lollygagging.

The train delivers me into the heart of old Copenhagen, a wonderful conglomeration of historic buildings and cobblestoned streets, some cordoned off for pedestrians only. The juxtaposition of the modern transportation system against the storied architecture is a little jarring. I'm here to see Peter Gundelach, a professor in the department of sociology at the University of Copenhagen. He has been Denmark's representative in putting together the European Values Survey for more than twenty years and is something of an expert on Danish happiness.

Gundelach, thoroughly academic in his speech, sidesteps the question of whether he himself is happy. Answering it would be untoward, he almost quips. Instead, he gives me a quick history lesson. Denmark's most recent big event, he says, happened in 1864, when Germany conquered about a third of its territory. Some of that land was regained in the twentieth century, but the traumatic experience helped unify Danes and shape their expectations in life. "People were able to adjust to this defeat," he says. "The defeat meant that Denmark had to have small expectations on what it could achieve internationally. They managed to build a feeling of [being] one people that could manage serious circumstances."[9]

Happiness levels are linked to such big, catastrophic events, which other countries have experienced more recently. It took Germans and other Europeans decades to get over World War II, while many Americans are still living in the psychological shadow of the 9/11 attacks. If there's a bright side to such tragic and traumatic events, it's that they can ultimately unify people and lead to higher happiness levels, but the effects often take time to coalesce. Countries where catastrophes don't generally happen tend to be happier than those that routinely experience war, terrorism, and other such trauma. Again, stability and economic prosperity rear their heads as all-important factors.

Gundelach's point, however, is that people everywhere—not just Danes—need to manage their expectations in life if they want to be happy. The prototypical "American Dream" encourages people to dream big and want the world ("and everything in it," as Tony Montana says in *Scarface*), but they're more likely to achieve satisfaction if they set modest goals. This might be hard to do in the current climate, where the media continually and subtly encourage people toward more. Britney Spears literally sings, "Gimme More"—but we'll get back to that shortly.

The other secret to Denmark's happiness, Gundelach explains, is that the government is integrated into almost every aspect of people's lives and beneficently so. My friends in Copenhagen frequently regale me with tales of the benefits and advantages the state makes possible: People get paid to go to university, rather than the other way around; taxes on fatty foods are redistributed to make organically grown goods cheaper; generous maternity benefits are supplemented by psychologist house calls to "see how you're doing," and so on. Much of this sounds like insane, interventionist socialism to my market-forces-hardened North American ear, but it's hard to argue with success. The Danes are a happy bunch.

The small population, in conjunction with the highly integrated government, also results in one strong positive happiness effect: trust. In that way, Denmark is like a small town. "It's a small homogeneous society where people know each other," Gundelach says. "They almost consider the state as a friend, compared to southern Europe, where it's considered an enemy."

That adds credence to the notion that growing individualism may be negatively affecting happiness, and it's certainly the case for loners in Denmark. It's a great place if you go with the crowd, but swimming against the stream can be tough. This is perhaps why we shouldn't trust international happiness surveys too much. "It's a way for people to tell that you are integrated into society more than it is a measure of happiness in a general philosophical sense," Gundelach says.

Alexander Kjerulf seconds that notion. His background is IT, but he quit that unfulfilling career path to become a unique sort of human resources consultant. He bills himself as a "chief happiness officer" and advises companies on how they can improve their employees' morale and work satisfaction. He comes at the issue of technology and happiness from an intriguing perspective: He's a big fan of progress, but he's also aware of the negatives it can create, especially its tendency to alienate people.

One of the big problems with workplaces today, he says, is that employees tend to spend their breaks browsing Facebook. In the past they used to talk to their colleagues while having a cigarette or coffee. People are less likely to talk to one another while out in public because typically they've glued their eyes to a smartphone. We also use these devices to navigate foreign cities and translate signs into our native tongues, whereas before we might have asked strangers for help. "That's too bad because that can make your whole vacation, that one stranger who showed you the way or bought you a cup of coffee," Kjerulf says. "That doesn't happen as much now."[10]

His point rings true. I switch off my phone's roaming capability while in Copenhagen because doing otherwise would cost a fortune. The upside? It forces me to ask a passerby for directions to the cafe in which Kjerulf and I met. I don't have any life-altering experiences along the way, but I also can't remember the last time I approached a stranger and initiated a conversation.

"We've taken away a lot of the things that make us unhappy, but we've forgotten a lot of things that make us happy," Kjerulf says, echoing the Venerable Noh Yu.

Still, there is that plus side, and Denmark is a big beneficiary. One of the biggest contributing factors to the country's overall life satisfaction levels is Flexicurity, an ingenious employment system that benefits every stakeholder in the country's economy. "If your job sucks, you can actually quit," Kjerulf says. "You don't necessarily need to wait until you find another job. You can say, 'Screw this.'" For workers, the government offers some of the most generous unemployment insurance benefits in the world—up to 80 percent of an individual's former salary, paid out for up to two years.

Employers also get considerable benefits from the Flexicurity system. It's relatively easy for them to fire someone if they can show that the person is underperforming or if the economy takes a turn for the worse. Knowing that someone can be cut easily makes companies less hesitant to bring them aboard in the first place and improves the hiring process. That's in stark contrast to the typical North American way of doing things, where hiring and firing are both laborious processes for employers and employees alike. It's even worse in France and Spain where overreaching labor laws often prevent companies from firing even the worst employees.

Finally, the third wedge of the Flexicurity model—the Danish government—uses a good chunk of high personal income taxes to fund retraining programs, thereby providing for people who are looking to change careers.

A person's job—what people spend most of their lives doing—obviously has a major effect on overall life satisfaction. In Denmark, people generally don't work at jobs they hate. If only that were true in North America, where large swaths of the population toil at jobs they despise with co-workers and bosses they loathe. Danes, on the other hand, generally don't hesitate to quit their jobs, train for new ones they like more, hold out until something better comes along, or even start their own businesses because a safety net allows them to do so and because, as we saw in chapter four, technology also enables them to do so. All of these factors together do much to explain why the country ranks atop entrepreneurship rankings. Not surprisingly, the European Commission is trying to implement the Flexicurity model in various forms across the continent.

The plumbing that underlies Denmark's labor system is, of course, technology. Like South Korea, Denmark is one of the most wired and wireless countries in the world. Its land-based and cellular Internet connectivity rank among the fastest and cheapest available. My friends' home Internet connection is the fastest I've ever seen and it costs less than a third of what I pay back home. Smartphone service, meanwhile, is among the cheapest in Europe—an amazing feat in a country where virtually everything is expensive. As such, everyone is well connected regardless of whether they live in big cities or small towns, making Denmark one of the few countries in the world with no digital divide between rich and poor. Enlightened social policy works hand-in-hand with technology to create an egalitarian society that offers numerous possibilities. Starting a business, taking retraining courses, and accessing government services are all cheap and easy.

Still, it's not all rainbows and roses. Kjerulf worries that Danes are succumbing to that hedonic treadmill. "The American dream used to be a small house with a bedroom that your two kids could share. Now, it's *MTV Cribs*," he says. "Danes are becoming more like Americans in this sense." One of the biggest and most ironic negatives of technology, particularly the communications kind, is that it expands people's awareness of what else is out there. The positive effects are obvious, however, as seen in recent years in the Middle East. In countries such as Egypt and Iran, connected citizens have risen up to demand more from their leaders after seeing the freedoms and wealth enjoyed by people in other parts of the world. The more people know about what their neighbors have, the more they tend to want it—and more of it.

That Danish trait of managed expectations that Gundelach mentioned is facing pressure worldwide. People everywhere are seeing everything that everyone else has. Rising conspicuous consumption, materialism, and greed have become the new danger. The phenomenon may be even worse or more powerful in a highly connected country such as Denmark.

The role that media plays in an individual's relative happiness can't be understated. A steadily increasing stream of advertising for an exponentially increasing stream of new products is creating wants and needs we never had before. Take the iPad, for example. It's fantastic and I can't imagine my life without it now. But has it made us any happier? That's the big question. The Buddhists say no, and maybe they're right.

Then there's the news media, continually telling us that the world is going to hell. It's become a cliché. Go to any news website or flip on the evening report on TV. All you're likely to read or see is what's going wrong in the world: death, corruption, crime, scandal, suffering—and, oh yes, maybe a story about a cute puppy to balance all that horror. The arrival and maturation of the Internet has kicked media—and our exposure to it—into the exponential. With so much of it out there and so much of it negative, that can't be good for us.

Tor Nørretranders, a veteran science journalist and author, has some unique views on how technology and the media interrelate with happiness, particularly as it pertains to Denmark. He believes the news media functions as an extension of our biological selves in terms of how it approaches what's going on in the world. Positive news—such as the world's rapid rise out of extreme poverty, discussed in chapter two—goes largely unreported, but the smallest negative event gets a lot of coverage because we're hard-wired for that sort of information. In that way, the media is our collective nervous system, delivering only the data pertinent to continued positive operation. "It's like your body: you get bad news from it in the form of pain, but you don't get happy news that your liver is actually functioning as normal," he says. "You don't get that because you don't want to pay attention to being happy all the time."[11]

The problem with this hard-wiring is that it creates a cloud of psychological negativity that can be difficult to escape and which can actually eat away at our happiness. The problem exacerbates when it comes to technology. The media loves stories about the possibility of Terminator-like robots running amok or about humans losing jobs to them, but the positive aspects of machine automation—how it leads to better, smarter jobs for humans—are rarely if ever discussed. The result is that people become fearful or distrustful of new technology.

Nørretranders worked at a science journal back in the 1970s, when the predominant public view on technological issues was that nuclear power was bad and IT stole human jobs. The publication's staff took a different perspective and focused on the positive effects of those issues. They framed the growth of IT and computers, for example, as a turning point for society—which quickly earned them the enmity of left-leaning thinkers, of which there are many in Denmark. The din of disagreement swallowed their point about how IT could make society enormously more democratic or potentially even more hierarchical. "There's this constant discussion over side effects instead of the effects of technology," he says. "It's easy to discuss. It's not intellectually or morally as demanding to discuss. We have to talk about where we want to take [technology]. That is the discussion."

Nørretranders also worries that media negativity is affecting the all-important trust factor in happiness. Trust really is Denmark's secret sauce, he says, because people don't worry about danger, which ultimately lowers transaction costs for the economy and contributes to the efficiency of the system. Lawyers don't have to be brought into every situation, and even rich people believe in taxes and charity since they understand the benefits those institutions deliver to them. A well-functioning public transit system, for example, ensures that employees get to work on time and that they're not miserable when they get there, which ultimately pays dividends to the business owner. In the United States, it's no coincidence that technology companies are leading the way with this sort of thinking. Many have instituted happiness-building measures for employees, including on-site day-care centers, gyms, and even movie theaters.

My conversation with Nørretranders eventually loops back to the game theory discussed back in chapter two. Despite the ever-present danger of greed and materialism growing, Danish people over the course of decades have come to understand the value of collaboration, sharing, and equality on a deep psychological level. Denmark's finely tuned trust system, although increasingly susceptible to American-style materialism, means that everyone understands the benefits of everyone else benefiting from the country's wealth. It's a truth the rest of the world is slowly discovering, Nørretranders says. "It's not to be romantic or nice to people, it's just more rational to do it that way."

## NOTHING GROSS ABOUT HAPPINESS

The future of happiness may not lie in pharmaceutical or even neuroscientific breakthroughs but rather in analytics. It's a fancy word also known as "Big Data," but it's really just a modern application of an age-old concept. Governments have been taking census polls of their subjects since Blarg and his compatriots picked a tribe leader. Census results are invaluable tools for setting policy because they give governments clarity on who they're governing, where those people are, and what their needs might be.

Census-inspired policies are generally aimed at boosting the public's happiness, but no government has gone as far in this regard as Bhutan's. The tiny kingdom of just 740,000 people boasts a similarly small GDP per capita of only about twenty-four hundred dollars. Its main businesses are agriculture and selling electricity to India, its huge neighbor to the south. Yet Bhutan long ago realized that money can't buy happiness.

As far back as 1729, the country's legal code declared that, "if the government cannot create happiness for its people, there is no purpose for the government to exist." In 1972, Fourth King Jigme Singye Wangchuck declared "gross national happiness" (GNH) more important than GDP. In 2008, Bhutan added that line of thinking to its constitution:

> *Gross National Happiness measures the quality of a country in a more holistic way and believes that the beneficial develop-ment of human society takes place when material and spiritual development occurs side by side to complement and reinforce each other.*[12]

More touchy-feely nonsense, right? Nope. The Bhutanese government isn't just talking, it's doing. It conducted its first national happiness survey in 2006, then again in 2008 and 2010. In each case, the questions asked were assessed and improved, the government calling the whole exercise a "living experiment." The 2010 survey consisted of nine domains further subdivided into thirty-three indicators that measured pretty much every aspect of a person's life, including assets and housing, household income, hours spent working versus sleeping, faith in government and political participation, health and spirituality levels, time and money donated to charity, and caring for the environment.

The ongoing survey provides an alternate measure of development by letting the government specifically target areas that need improvement. Officials can see that one region of the country may be lagging in essential services such as hospitals, for example, or that people in another area aren't getting enough sleep for one reason or another. Then they can deploy resources to fix the issues.

It's micro-management at its finest and also state-oriented—a very Danish approach—in that the government decides the factors that make people happy. But the survey and its weighting of indicators purposely contains some leeway to account for personal choice. Literacy level is one indicator, but a citizen can be considered quite happy without knowing how to read by scoring well in other categories.

This is a low-tech version of Big Data in action. The same exercise, applied with online analytics in a more technologically advanced nation, can provide governments similarly fine-grained information with which to set policy and deploy resources. Many countries already are doing that in a variety of ways, from managing electricity grids and police forces to urban planning and education spending. There are privacy issues, of course, which we discussed in chapter seven, but Big Data holds the promise of improving everyone's happiness levels by enabling the better deployment of state resources.

So how happy are the Bhutanese? About a tenth of the population scored under the 50 percent mark in the 2010 survey, meaning they were considered unhappy. Close to half scored between 50 and 65 percent, or the narrowly happy category. A third rated extensively happy with a score between 66 and 76 percent, with the remaining 8 percent of the population rating "deeply happy." Close to 60 percent of people are considered to be "not yet happy," which means that Bhutan still has a long way to go. The country does relatively well in international surveys, but its own internal measures are obviously more strict.

The funny thing about Bhutan's GNH is that it's somewhat hypocritical. While striving for the admirable goal of boosting the happiness of its people, the country also has one of the highest levels of refugees per capita, with some observers claiming that up to a sixth of the population has sought asylum in other countries since 1991.[13] The refugees consist mainly of the Lhotshampas, people of Nepalese origins discriminated against and encouraged to leave out of fear that they will overwhelm the native Bhutanese population. The government aims to promote happiness—but only for a select group of people. Bhutan's GNP also proves that, despite all the touchy-feely nonsense, a country still needs economic prosperity and safety as a baseline for other happiness endeavors.

Nevertheless, the ideal behind Gross National Happiness is good. The goals of strong family structures, respect for culture and traditions, a clean environment, and peaceful coexistence with others all lie at the heart of Bhutan's efforts, if not always in its actions. As Fifth King Jigme Khesar Namgyel Wangchuck said in a 2009 lecture, "The duty of our government must be to ensure that these invaluable elements contributing to the happiness and well-being of our people are nurtured and protected. Our government must be human."[14]

## SUNGLASSES FOR THE FUTURE

The governments of Denmark and Bhutan have it mostly right. Closer relationships, trust, altruism, equality, and freedom matter most to happiness in a post-prosperous world. Technology is helping with some of these factors and harming with others. It's key to keep in mind, during a period of exponential and ubiquitous technological growth, that we must identify both the positive and negative effects so that the good can be accentuated and the bad ameliorated. Above all, people and governments must realize—or be reminded—that happiness should be the ultimate goal of society.

So far, many of the world's developed nations, especially the United States, have focused mainly on the bottom half of

Maslow's hierarchy of needs, the basic physiological and economic necessities. The rest of the world, as we saw in earlier chapters, is benefiting from this focus. Millions in developing nations are rising from mere subsistence into a life that offers options and even potential happiness. It's happening because of a global realization of advanced nations' best economic practices. Driven by technology, open markets and inter-operation are the best ways known to facilitate the fulfillment of those particular needs. As more countries join the world market, humanity will speed ever faster to a future in which no one needs to despair and everyone has hope.

But globalization is a two-way street—or rather a multi-way intersection—where the best practices of different societies can and will rub off on others. Happiness studies and Big Data are invaluable resources and tools with which all nations can learn about the more advanced stages of Maslow's hierarchy and what effects higher levels of happiness. Countries such as the United States and the United Kingdom, where happiness has stagnated recently, can learn from the efforts of places such as Denmark or Bhutan and figure out how to emulate them within their own systems. American-style corporatism may be bringing the world to a certain level of prosperity, but other countries can help take humanity to the next level.

In that sense, we stand at a novel point in history. As noted earlier in this chapter, there hasn't been a way to tell how happy people have been historically because we've started only recently to keep track in any large-scale way. Now that we're paying attention and developing the tools, we can form best happiness practices not based on deceptive economic trickle-down theories. That's a big reason to be optimistic about the future because it's going to be a lot happier than today.

# 10

# Conclusion: Marx Was Right (Sort Of)

*If no mistake have you made, yet losing you are, a different game you should play.*

—YODA

When we first met John Seely Brown in chapter one, he was cruising down the California coastline in his high-tech Porsche. But he's not always doing that. Most of the time he's spreading confusion—literally. He calls himself the "chief of confusion." He's a technology consultant to businesses and organizations, but his role isn't to provide solutions to issues. Rather, like Socrates, he tries to help people ask the right questions. "In this new world we're living in, questions may be more powerful than answers," he says.[1]

More Yoda-isms—appropriate for one of the men who coined the term "ubiquitous computing," a concept about how technology invisibly surrounds us. It sounds like the Force, but I've come to understand his meaning over the past few years working on this book, which has been predicated on questions, starting with a very basic one: What is technology doing to us? In each chapter, we've extrapolated that fundamental question into the various aspects of human existence: How is technology affecting our health, our relationships, our collective happiness, and so on?

As the religious adherents we met in chapter eight pointed out, any conclusions that we may draw inevitably lead to more questions. In figuring out what this all means and how to apply this knowledge, we have to decide what to ask ourselves going forward. We know what technology is doing to us and has done over several millennia of evolution, but what do we do about it now?

Do we like its effects? If not, is there any way to stop or reverse them? What will Humans 4.0 look like?

In the broadest sense, technology's effects on the world and humanity generally have been benevolent. Advancing technology, along with the globalization it enables and deepens, has driven the world economy for centuries by creating possibilities and efficiencies, allowing new opportunities, bridging distances, and equalizing positions. Governments that have allowed or encouraged these two interlinked forces to take root and expand freely have benefited tremendously; those that haven't find themselves and their people left behind. With bottom-up revolutions continuing to foment and explode in the Middle East, it's clear that people in some of these outlier nations are no longer content with taking a back seat to history's direction. We can only hope that for the people in willfully oppressed parts of the world, such as North Korea, technologically driven globalism seeps through the cracks and works to bring them into the fold.

Despite short-term blips such as Occupy Wall Street, the long-term global trend is evident: The world has been heading toward greater prosperity for some time, with particular acceleration in the past few centuries. Not just is computing becoming ubiquitous; so too—one day—will relative wealth be everywhere.

This richness translates into many measurable benefits, starting simply with how long people live. With basic human needs being met and technology providing cures and treatments for all manner of ailments and previously fatal diseases, children will grow into adults who then achieve life spans comparable to tortoises. Their long lives will teem with comforts, from automobiles and air conditioners to electric screwdrivers and robot vacuums. Life will no longer be as "nasty, brutish and short" as it was or even as it is—except, of course, when it comes to cleaning the toilet.

This important improvement in the basics and some of the luxuries shouldn't be understated, but unfortunately it is. In the advanced world, we generally don't think about how good we have it. We rarely experience the early death of a child, and that's increasingly applying to adults as well. We almost never contemplate that, just a few short centuries ago, families didn't have electricity and had to sleep together to keep warm. Happiness baselines, so we forget how hard life can be and instead complain about how our Wi-Fi range doesn't cover the backyard or that we can't use the bathroom because the maid is cleaning it. Perhaps a biological imperative paradoxically prompts us to focus on the negatives, but it's a shame nevertheless how little we consider how far we've come and how much we fret about what really doesn't matter.

The big-picture advancements have also resulted in one of humanity's greatest and again unheralded achievements: the decline of war. The conditions that drive people to fight in conflicts and commit terrorism still exist in many parts of the world, but the rapid spread of prosperity has diminished those causal factors significantly and will continue to do so. More and more people are gaining things they'd rather not lose—family and loved ones, possessions, or even just hope—which is something that can happen when they engage in violence against their neighbors. At this point, game theory and the Prisoner's Dilemma become more relevant to the average person: The creep of prosperity tells individuals that everyone gains by cooperation, or, more specifically, that no one loses by it. We'll return to this important development shortly.

Factors other than economics lead people to harm their neighbors, but here too technology is helping. Advances in neuroscience are giving medical practitioners a better understanding of the mind and personality, so perhaps even the unwell people who shoot up schools or commit other heinous murders might receive improved diagnoses and treatments before they do so. With serious efforts and resources going into understanding the brain—the real "final frontier"—we may eventually understand evil in all its forms, and how to prevent it. This progress, combined with prosperity's mitigating effects on violence, means that progress is pointing to a decline in conflict of all sorts. The idealistic notion of entirely eliminating people's desires to kill or inflict harm against their neighbors doesn't seem so fanciful when cast in that light.

Greater prosperity, better disease treatment, and the decline of violence all add up to more people enjoying lives unprecedented in length—and they're getting longer all the time. The natural fear that follows is that this is ultimately contributing to overpopulation and that a shortage of resources, particularly food, is therefore inevitable. But that's not the case. As we've seen, one of the other major effects of income growth is a corresponding decline in the number of children born. That rate is falling dramatically and considerably faster than the death rate around the world, which is why the population will plateau if not dip within this century.

Technological advancement in food production shows no sign of abating. In fact, it may reach unprecedented growth as genetically modified organisms gain more acceptance and new characteristics. Bigger fish, insect-resistant crops, and plants needing little water are all near-term technologies currently in development. Regulators, the public, and even individuals who previously criticized them, meanwhile, are becoming less skeptical. Beyond genetic engineering, breakthroughs in creating synthetic meat have allowed scientists to grow beef and the like in labs from animal embryos. In 2013, a synthetic hamburger—bankrolled by Google co-founder Sergey Brin—received positive reviews from taste testers, paving the way for feeding the newly emergent middle classes of India, China, and other rapidly developing nations. Such breakthroughs hold the promise of feeding billions of people cheaply, while at the same time addressing animal rights concerns. Imagine: Hindus eating burgers! For decades the media has perpetuated the Malthusian nightmare of widespread food shortages caused by overpopulation, but the reality is quite the opposite when demographic trends combine with ongoing technological advances. With a plateauing population, it's entirely possible the world will experience an overabundance of food this century. A hundred years from now, it may be cheap and plentiful.

Of course, not all the large-scale effects of advancing technology have been good. The worst of it has been the toll on the environment. As people move out of poverty, they acquire cars, air conditioners, and flat-screen televisions, all of which use energy and create waste. The threat of ruining the planet through prosperity is real and being exacerbated by that same emergent middle class, which means we must take steps to avoid this fate before it's too late.

Fortunately, many efforts are under way to mitigate the paradoxical effects of wealth, some more complex than others. My favorite is a low-tech solution for stopping hurricanes thought up by former Microsoft chief technology officer Nathan Myhrvold and explained in Steven Levitt's *SuperFreakonomics*. The idea is to float large rings in the ocean with equally giant plastic cylinders attached to their undersides. The cylinders would force colder water to the surface to alleviate the warm-water conditions that cause hurricanes to form. It's a goofy concept that may never come to pass, but it's just one example of the sort of ingenuity—and creative entrepreneurialism—at work around the world. Solutions to prosperity-induced environmental damage will emerge . . . hopefully before giant disasters like the Boxing Day tsunami and Hurricane Katrina hit again.

## AGREE TO DISAGREE

On a more personal level, technology's effects are more mixed. As we've seen over the second half of this book, technology is enabling and accelerating individualism on every level: the jobs and companies we create and where we work, the art and entertainment we create and consume, the relationships we have with other people and organizations. Whether it's the photos, books, blogs, and music we make and share or the new business ventures we attempt, technology is allowing us to be more creative. It's also lessening our dependence on social institutions such as marriage and religion by allowing us to seek out and form our own social circles more tailored to our own interests. Social relationships now are more likely to be grassroots than imposed from the top down by institutions.

In each case, the individual is being empowered to fulfill the upper echelon of Maslow's hierarchy of needs. This is great because, with the basic requirements of life met, we can search out the deeper sources of happiness and satisfaction: who we are, what greater purpose we have, and how we achieve it. In many ways, technology is empowering people to discover their true identities and reasons for being, which means that real-world evolution is playing out very differently from how some dystopian science fiction predicted. Technology hasn't absorbed us all into a hive mind, like the Borg on *Star Trek*. Quite the opposite, actually. It might be tough to swallow for the average person working a soul-crushing job in an office somewhere, but it's not society lacking in this case; it's that the individual hasn't yet awakened to the possibilities that society is making available to him or her, either on a personal or career level.

The larger growth of expression and commerce says something fundamentally poignant about humans in general: We're a creative bunch, with no apparent limits on that creativity. Just when we think we've run out of artistic or commercial ideas and the movie theaters are full of nothing but sequels, along comes Nirvana, Amazon.com, or *Breaking Bad* to change everything. New creation arises that takes what was done before and reconfigures it into something new and often better. In that sense, we're more than capable of keeping up with exponentiality because of our unceasing ability to think combinatorily. The old cliché holds that nothing is certain but death and taxes—and maybe not much longer for the former—but we can add human combinatorial creativity to that list. It's inevitable and will continue.

The growth of individualism will also continue and may increase dramatically. The next frontiers in technology—robotics, nanotech, neuroscience, and biotech—promise huge improvements in this area. Genome sequencing becomes significantly cheaper every year. Soon we'll be able to map and read our entire genetic code through an app on our phone while nano-sensors inside our bodies report on troubles as they arise. As Eric Topol, the digital doctor we met in chapter three explained, this will lead to an explosion in personalized health care that eliminates much of the guesswork about our individual issues. Combine that with the brain map currently being researched and possibly even a personality map that follows, and, as per the directive of the Delphic Oracle, we will know ourselves completely, inside and out. That makes the strides in individualism we've made so far seem quaint.

The flip side of this coin, however, is insularity, a less desirable trait because of the strong evidence linking it to unhappiness. As we saw, fewer group and organizational bonds—such as fewer people getting married and joining community-minded institutions like churches—are leading to growing insularity, which can cause distrust among people. Greater individualism and insularity can also cause divisiveness to grow. That doesn't necessarily mean violent war, but anyone who's watched a heated discussion online transform into a vitriolic argument and then all-out flame war knows that the distinction might not be all that great. With individuals increasingly enabled to find their respective voices, there's a greater potential for disagreement.

People are even questioning physical facts, as Suzanne LaBarre, online content director of *Popular Science,* lamented in 2013 when the magazine shut down reader comments on stories. "Everything, from evolution to the origins of climate change, is mistakenly up for grabs again. Scientific certainty is just another thing for two people to 'debate' on television," she wrote in an editorial explaining the publication's decision.[2] Some might consider that debate a good thing since communication and conversation often deepen our understanding of a topic, but I get a sinking feeling in my stomach when even basic facts fall victim to our divisiveness.

The preponderance of minority governments in many advanced nations both reflects this growing individualism and highlights the need to adjust to it. It's easy, after all, to govern people with uniform views who agree with one another, but how will we do so when that's no longer the case? Bhutan stands as an unfortunate example. Even a tiny country focused on universal happiness can't please all of its people. In the face of this growing individualism, fundamental changes to governmental structures and even borders are inevitable. We could see the rise of micro-nations, consisting of small groups of people who agree with one another. Perhaps the long-held notion that democracy is the best form of government will face serious challenges. These are the questions we need to ask.

Fortunately, our disagreements don't look like they'll manifest themselves violently, online comments notwithstanding. Exceptions will occur, of course, but for the most part people have subconsciously—maybe even consciously—realized that they're living the Prisoner's Dilemma, in which the lines between self-enrichment and universal harmony have blurred. That brings us to the heart of what I think Humans 3.0 really are.

## CAVE-MOUNTED TELEVISION

At first glance, the directionality of the world and its people seems to be diverging. Things are getting better in the big picture, but that isn't necessarily the case when we drill further down. Technology is bringing us to a certain level of comfort, but then it stalls out, if it doesn't wind up doing more harm than good in the end. The question, then, is clear: What good is further progress if the links between us are fraying and we're not getting any happier?

Ultimately, it's not that simple. What we're experiencing is a technologically driven social dialectic. The first stage means the splitting of human development into two paradoxically divergent paths—globalized harmony versus rampant individualism—but such dialectics inevitably converge. Georg Hegel, drawing on his forerunner Immanuel Kant, postulated that history was moving in such a dialectical pattern. In perhaps his most well-known example, he suggested that people naturally divide into two classes: masters and slaves. But with masters inevitably becoming dependent on the slaves, the slaves eventually gain their freedom.

Karl Marx took the idea further with his belief that human evolution represented the dialectical story of class struggle. He thought human history would have five epochs, starting with the sort of primitive communism that our caveman friend Blarg lived, in which no one owned anything and only the common good mattered. From there, mankind evolved into the master-slave relationship common in much of the ancient world, followed by the lord-and-serf dichotomy of feudalism in the Middle Ages. Capitalism followed, where property owners or the bourgeois lorded over workers or proletariats and formed the system through which our modern era operates. In each previous case, the oppressed class rose up to replace the ruling class, resulting in a short-term Hegelian synthesis. But that synthesis ultimately ushered in a new divergent dialectic. The final synthesis, Marx theorized, was a return to Blarg-ian communism. With people tired of oppressing their neighbors for millennia, a new era of collegiality—where again, only communal harmony mattered—would arise.

Where Marx and his many adherents, from Vladimir Lenin to Mao Tse-tung, got it horribly wrong was believing that they could force this ultimate synthesis, that political parties or even charismatic individuals could somehow manufacture this ideal communism and foist it upon the masses. As the fall of this imposed communism in the twentieth century proved—dramatically in the Soviet Union and more subtly in China—it just isn't possible. People aren't willing to give up everything they have because some ideology tells them to do so, especially when the systems ended up as oppressive and repressive corruptions of the ideal that inspired them in the first place. Worse still, communism wasn't possible unless everyone went along for the ride. People living behind the Iron Curtain weren't willing to accept less freedom or material wealth when they easily could see how much of both the people on the other side had.

But in the long term, Marx wasn't entirely wrong. Communism may be humanity's ultimate dialectical synthesis. It's just happening naturally and taking a different form than he or his followers anticipated. The key to it all is the complexity of human understanding.

Consider that flat-panel TVs and genetic engineering were inconceivable to Blarg; there were too many intermediary steps to take before he could even imagine them. He needed to learn how to walk before he could dream of running. Yet, with every discovery, invention, and innovation since then, the human brain has become more capable of conceiving this figurative "running" —up to a point, of course. Going back to the future prediction discussed in chapter one, it may not be difficult at all to envision where the world is going over the next few decades. It's only after the Singularity, which observers believe will happen in the middle of this century, that the future gets hazy. It's difficult to imagine what might happen after computers surpass our collective intelligence, which explains why so many people are afraid of this inevitability.

Still, thoughts that were overly complex just years ago—even to the average person—are becoming simpler and clearer by the day, which proves that it isn't just technology that is advancing exponentially. So too is our ability to think in complex terms . . . reality television and Justin Bieber notwithstanding. This ability to think bigger and more deeply is manifesting itself both externally, as we expand our thoughts into the outer universe, and internally. Aside from research taking place in biotechnology, genetics, and neuroscience, we're discovering new depths to our selves and our creativity. Thus technological advance is furthering what we understand and know about ourselves.

These advances give us a different way of looking at the Prisoner's Dilemma and game theory in general, both of which can test the ability to think in simple or complex terms. Choosing the path of self-enrichment at the expense of others has always been the simplistic course of action; gaining less without costing someone else in the process has been the more complicated option, requiring the chess-like ability to think several moves ahead. The history of mankind has been a tale of bouncing between these two extremes, but we're becoming more capable of understanding the second path as our thinking gets more complex. We are starting to realize that we benefit individually when everyone gains.

That's a hard pill to swallow given the constant, daily reminders we have of basic human greed. If a politician isn't embezzling public funds, a big company is lobbying to crush environmental regulations or prohibit municipalities from building their own telecommunications networks. Greed, in all its sad and sickening incarnations, affects everyone.

Yet the scientific evidence linking greed to a biological imperative is scant, which means it's largely a socialized condition. In other words, we aren't born with greed, we learn it. In *Nonzero: The Logic of Human Destiny*, journalist Robert Wright argues that history has been the story of people increasingly learning that selfless cooperation is the best way to get what they selfishly want. From North American natives to their counterparts on the other side of the world in the "inscrutable Orient," people historically have shown a tendency to help one another with the knowledge—either conscious or subconscious—that such altruism will benefit themselves either directly or indirectly. It may have taken the form of one tribe giving surplus supplies to another or one country giving economic aid to another, but there's a deep, fundamental understanding of the benefits of this sort of sharing. Globalization has "been in the cards not just since the invention of the telegraph or the steamship, or even the written word or the wheel, but since the invention of life," says Wright.[3] "All along, the relentless logic of non-zero-sumness has been pointing toward this age in which relations among nations are growing more non-zero-sum year by year."

The opposite of greed, altruism, may in fact be a biological imperative not just for humans, but for all organisms. Consider the man who volunteered for watch duty to keep his Medieval village safe from pillaging invaders. This brave and selfless sentry stood a higher chance of dying than those he was guarding, so wouldn't natural selection weed out such foolhardy individuals? Wouldn't the selfish souls who hid in their well-appointed fortifications have won out, genetically speaking? Evolutionary psychology tells us otherwise, as we'll see in a moment. Countless other examples of selflessness permeate daily life, despite the well-publicized stories of shameless greed and corruption.

The concept is known as kin selection, sometimes referred to as Hamilton's Rule for W. D. Hamilton, the evolutionary biologist who popularized it in the sixties. The mathematical explanation is complicated, but he theorized that humans perform altruistic acts if they can identify a kinship with their beneficiary. In other words, the presence of the trait that prompted the courageous sentry to risk his life enabled more people in his overall kin group to survive more often and reproduce. Under that definition, we naturally are more altruistic with close friends and family. Perhaps that's why we give them gifts on birthdays or at Christmas. It also explains why certain monkeys and shrimp engage in allomothering, caring for the offspring of their relatives. Researchers proved Hamilton's Rule in 2010 when they found that red squirrels in the Yukon often adopted the orphaned offspring of their relatives, but not those of nonrelatives. "By focusing on adoption in an asocial species, our study provides a clear test of Hamilton's Rule that explains the persistence of occasional altruism in a natural mammal population," the researchers wrote.[4]

Even non-organics impulsively come to understand the benefits of cooperation and altruism. Swiss researchers conducted an experiment in biological evolution in 2011 that found swarms of tiny robots naturally gravitated toward this kind of behavior. The sugar-cube-sized robots began as directionless entities wandering aimlessly into walls. They eventually evolved into foragers that collected little tokens representing food, which the scientists expected. What surprised them, however, is that the machines continued to evolve to the point where they helped one another. One experiment featured large tokens that only several robots together could push, and, without that specific programming, the machines naturally cooperated to do just that.[5] The scientists concluded that altruism "should only evolve among related individuals, and this is also what has been found in a wide range of organisms, ranging from bacteria to social insects and social vertebrates."

We can extrapolate such findings, along with Hamilton's Rule, into the larger context of the world's evolution. All organisms have a natural tendency to cooperate with one another both because of inherent genetic links and because of logical self-interest. That tendency is often subsumed by the equally natural drive to compete when resources are scarce, but therein lies the answer. As resources become more plentiful, as the world marches onward to prosperity, the urge to compete lessens. Adding to that are the rapidly growing connections, inter-relatedness and inter-dependency that globalization causes. As one country comes to know and depend more on another, the motivations for altruism between the two inevitably increase.

This is also the case when it comes to people. If we're more motivated to give of ourselves to people close to us, this circle of closeness is going to expand as globalization continues. A hundred years ago, people in the West didn't have many ways in which to feel kinship with those in Africa or the Middle East. Now, thanks in large part to technology, increasingly they do. The same is inevitable in the reverse, as those parts of the world become better connected. Even the superficial connections brought about by social media help in this regard, since we're more likely to help out that "friend" we barely know on Facebook than we are a total stranger. In this sense, the social network's claims of shrinking degrees of separation carry some valid weight.

## TO INFINITY AND BEYOND

The diverging dialectic, then—globalized harmony versus rampant individualism—is destined for a new kind of synthesis. As in the Prisoner's Dilemma and unlike in Marx's models, it's a situation where both directions can win and merge, rather than one subsuming the other. In a simpler time, that may have been the necessary outcome, but with humanity's understanding of itself becoming more complex, it's possible to have both.

Let's go back to the example of the smartphone from the first chapter. That simple device derives from thousands of different individual technologies. Countless inventors and companies are competing to make the best cameras, GPS chips, or radio antennae that go into such devices. It's the same with humanity. Many individuals and countries are competing against one another in various aspects of life and economy. Yet together, those competing components form something greater than their parts.

That has always been the case, but we're only just now becoming aware of it thanks to an increasing ability to think in such increasingly complex terms. Back in chapter two, we saw a Price-WaterhouseCoopers report on how the world's emerging economies aren't necessarily bad news for established nations since everyone likely will benefit. Just as countries can specialize in their areas of competitive advantage yet still prosper all together because of it, so too can people. Harmony and individualism aren't engaged in a zero-sum game but a mutually beneficial evolution.

When I started working on this book, my hypothesis was that Humans 3.0 were the embodiment of the Anthropocene era, where mankind had evolved from being subjugated by nature to mastering it. Over the past few years of delving into this evolution, however, my views have evolved. Throughout history, we've bounced between competition and cooperation, often not aware that we were doing so. Humans 3.0, the updated software to the hardware supplied by nature, isn't just the synthesis of that dialectic; it also represents the emergence of our consciousness of it. We're not just becoming a different kind of people; we're becoming aware of the fact that we're becoming a different kind of people. Humans 3.0 therefore aren't just group harmony or competitive individualism. We're both, a combination of the two, an unprecedented step in history. It's been a slow process so far to get here, but, just like the technology behind it, it's speeding up.

Becoming more optimistic about the future and technology's effects on humanity than when I started this book surprised me. Initially I believed that technology was a benevolent force overall because its positive effects outweighed its negative effects, but that's a simplistic outlook. It's not just about whether the possibilities created by smartphones are more important or widespread than some of the addictive behaviors they've propagated; it's about gaining a deeper insight into what technology says about us and what it enables on a more soulful level. My own consideration of this book's central thesis has become more complex.

As I get older, I'm getting younger in a sense. I play more video games now than I ever did as a child and I have more Legos now than I ever did too. I recently took up volleyball again, and just a few years ago I developed a taste for curry, a spice I couldn't tolerate when I was younger. I also started listening to Rush, the legendary Canadian rock trio. As a teenager, I'm not sure which I hated more—the spice, or the band. Now, Rush is the most sublime music I can contemplate.

I like to think my horizons are expanding and I'm devoting more time to leisurely—some would say childish—pursuits because I can. At the risk of hyperbolizing, much of that is because technology has made me prosperous. I make my living writing about it. I'm far from rich, but it enables my career and lifestyle. I can explore and discover more of Maslow's hierarchy as it pertains to my own situation. However, like most self-employed people who spend inordinate amounts of time working at home, alone, I'm more cognizant of the disconnective side effects than most—particularly if the growing number of conversations I'm having with my cats offers any indication. Fortunately, I'm aware that my forced absence from a social workplace is having adverse effects on my sanity, which is perhaps why I've started purposely partaking in group activities like sports. I just don't appreciate technology's positive effects on the world and humanity, I'm grateful for what it's done for me personally, and cognizant of where it may be harmful.

As for Humans 4.0, I can't wait to transfer my essence into a machine and fly across the galaxy. But we won't have to cross that bridge—or write that book—for at least another few decades.

# ACKNOWLEDGMENTS

If there is such a thing as Humans 4.0—an even more enlightened kind of people—I'd like to nominate the individuals who helped make this book possible.

A heartfelt thanks goes to James Jayo for being the first to see the potential of this idea and for allowing it to proceed, and for giving the manuscript such a superb edit. I'm also grateful to Susanne Alexander at Goose Lane Editions and Scott Pack at The Friday Project for their enthusiasm and support.

Jordan Kerbel was a fantastic facilitator in Israel while Avi Ben Josef was an excellent tour guide (and guitarist). David Carruth was immensely helpful in translating Korean for me, while my good friends Kasi Desfor and Gina Murray were exceptionally welcoming hosts in Denmark.

A big thank you goes to everyone who took the time to share their expertise with me for this project, but I'm afraid they're too numerous to list here. I hope that a sincere group "thanks" is enough, but if it isn't, please let me know and I'll buy you a beer.

My friends Kenny Yum, Andre Mayer, and Shane Dingman—all editors extraordinaire—each did me a huge favor by reading through early versions of the manuscript. Without their help, it surely would have been less readable and contained considerably more vulgar attempts at humor.

Throughout the whole process, I often marveled at the patience of my agents at Westwood Creative. How Chris Casuccio and John Pearce put up with me at times, I'll never know, but I'm glad they did.

Speaking of patience, my beautiful wife, Claudette, gets the biggest thanks of all for joining me on this rollercoaster. Writing a book, just like human evolution, is full of ups and downs. I'm immensely grateful that she stood by me through all of it.

Last but not least, a heartfelt thank you, gentle reader, for choosing to spend your time here.

# NOTES

## 1. EVOLUTION: OF RICE AND MEN

1. Ray Kurzweil, *The Age of Spiritual Machines* (New York: Penguin Books, 2000), 36–37.
2. Author's interview with John Seely Brown, June 2012.
3. The number is taken from two sources: "Smartphones in Use Surpass 1 Billion, Will Double by 2015," *Bloomberg*, October 17, 2012, www.bloomberg.com/news/2012-10-17/smartphones-in-use-surpass-1-billion-will-double-by-2015.html, and "Computers Sold in the World This Year," *Worldometers*, April 1, 2013, www.worldometers.info/computers/.
4. Author's interview with Alfred Spector, June 2012.
5. Author's interview with Bill Buxton, June 2012.
6. "Founder of Singularity University Talks About His Unusual New Institution," *Chronicle of Higher Education*, February 3, 2009, http://chronicle.com/blogs/wiredcampus/founder-of-singularity-university-talks-about-his-unusual-new-institution/4506.
7. Hans Moravec, "When Will Computer Hardware Match the Human Brain?," *Journal of Evolution and Technology 1*, 1998 (December 1997), www.transhumanist.com/volume1/moravec.htm.
8. Ibid.
9. "2045: The Year Man Becomes Immortal," *Time*, February 10, 2011, www.time.com/time/magazine/article/0,9171,2048299-1,00.html.
10. Ibid.
11. Author's interview with Steve Sasson, December 2009.
12. "Mathematicians Predict the Future with Data from the Past," *Wired*, April 10, 2013, www.wired.com/wiredenterprise/2013/04/cliodynamics-peter-turchin/all/.

## 2. ECONOMICS: WIDGETS ARE LIKE THE AVENGERS

1. Angus Maddison, *The World Economy* (Paris: OECD Publishing, 2006), 25.
2. Ibid., 103.
3. Ibid., 261.
4. Google public data: www.google.ca/publicdata/explore?ds=d5bncppjof8f9_&met_y=ny_gdp_mktp_cd&tdim=true&dl=en&hl=en&q=global+gdp. The International Monetary Fund estimates this to be higher, around $78 trillion, www.imf.org/external/pubs/ft/weo/2012/02/.
5. As of December 5, 2012.
6. World Bank, "World Bank Sees Progress Against Extreme Poverty, but Flags Vulnerabilities," February 29, 2012, http://web.worldbank.org/WBSITE/EXTERNAL/NEWS/0,,contentMDK:23130032~pagePK:64257043~piPK:437376~theSitePK:4607,00.html.

7. McKinsey Global Institute, "Internet Matters: The Net's Sweeping Impact on Growth, Jobs and Prosperity," May 2011, www.mckinsey.com/insights/mgi/research/technology_and_innovation/Internet_matters.

8. "Super Cycle Leaves No Economy Behind as Davos Shifts to Growth," *Bloomberg*, January 23, 2011, www.bloomberg.com/news/2011-01-23/super-cycle-leaves-no-economy-behind-as-davos-shifts-to-growth-from-crisis.html.

9. "Top 10 Largest Economies in 2020," *Euromonitor International*, July 7, 2010, http://blog.euromonitor.com/2010/07/special-report-top-10-largest-economies-in-2020.html.

10. "The Economic History of the Last 2,000 years in 1 Little Graph," *The Atlantic*, June 19, 2012, www.theatlantic.com/business/archive/2012/06/the-economic-history-of-the-last-2-000-years-in-1-little-graph/258676/.

11. PriceWaterhouseCoopers, "The World in 2050," PriceWaterhouseCoopers LLP, 2011, 24.

12. Ibid., 23.

13. "Super Cycle," *Bloomberg*.

14. CNET, "Bill Gates Predicts No Poor Countries by 2035," January 21, 2014, http://news.cnet.com/8301-1023_3-57617531-93/bill-gates-predicts-no-poor-countries-by-2035/.

15. "For Richer, for Poorer," *The Economist*, October 13, 2012, www.economist.com/node/21564414.

16. Ibid.

17. According to Box Office Mojo, http://boxofficemojo.com/movies/?id=avengers11.htm.

18. "Wealth of Billionaires Soars 14 percent Last Year to a Combined $6.2 Trillion While Everyone Else (Even Millionaires) Have Taken a Beating," *Daily Mail*, September 18, 2012, www.dailymail.co.uk/news/article-2204829/Wealth-billionaires-soars-14-percent-year-combined-6-2TRILLION.html.

19. "The Age of Inequality," *New Scientist*, July 2012, 44.

20. Ibid., 42.

21. "For Richer, for Poorer," *The Economist*.

22. "Prisoner's Dilemma," Stanford Encyclopedia of Philosophy, September 4, 1997, http://plato.stanford.edu/entries/prisoner-dilemma/.

23. Ernst Fehr and Urs Fischbacher, "The Nature of Human Altruism," *Nature*, issue 425, October 23, 2003, 785–91, and Amos Tversky, *Preference, Belief, and Similarity: Selected Writings* (Cambridge: MIT Press, 2003).

24. Wikipedia is a good place to start for lists of wars by death toll: http://en.wikipedia.org/wiki/List_of_wars_by_death_toll.

25. Estimate comes from "Afghanistan, Iraq Wars Killed 132,000 Civilians, Report Says," *Wired*, June 29, 2011, www.wired.com/dangerroom/2011/06/afghanistan-iraq-wars-killed-132000-civilians-report-says/, and iCasualty, http://icasualties.org/.

26. Compiled from Wikipedia, http://en.wikipedia.org/wiki/List_of_battles_and_other_violent_events_by_death_toll#Terrorist_attacks.

27. "The Number of Armed Conflicts Increased Strongly in 2011," Phys.org, July 13, 2012, http://phys.org/news/2012-07-armed-conflicts-strongly.html.

28. "Root Out Seeds of Terrorism in sub-Saharan Countries," *USA Today*, April 14, 2003, http://usatoday30.usatoday.com/news/opinion/columnist/wickham/2003-04-14-wickham_x.htm.

29. "The Age of Inequality," *New Scientist*, 44.

30. "Bacteria Use Chat to Play the 'Prisoner's Dilemma' Game in Deciding Their Fate," *Science Daily*, March 27, 2012, www.sciencedaily.com/releases/2012/03/120327215704.htm.

31. Maddison, *The World Economy*, 19.

## 3. HEALTH: THE UNBEARABLE VAMPIRENESS OF BEING

1. According to the American Human Development Project's *Measure of America 2013-2014*.

2. Author's interview with Anne Rice, March 2013.

3. Angus Maddison, *The World Economy* (Paris: OECD Publishing, 2006), 31.

4. National Institute of Population and Social Security Research, www.ipss.go.jp/pp-newest/e/ppfj02/t_6_e.html.

5. Maddison, *The World Economy*, 31.

6. United Nations, *World Population to 2300* (New York: United Nations, 2004), 25.

7. United Nations data can be found at: http://data.un.org/Data.aspx?q=world+population&d=PopDiv&f=variableID%3A53%3BcrID%3A900.

8. "History's Mysteries: Why Do Birth Rates Decrease When Societies Modernize?," *Psychology Today*, March 14, 2009, www.psychologytoday.com/blog/the-narcissus-in-all-us/200903/history-s-mysteries-why-do-birth-rates-decrease-when-societies-m.

9. "The Best Story in Development," *The Economist*, May 19, 2012, www.economist.com/node/21555571.

10. United Nations Department of Economic and Social Affairs, "Population Ageing and Development," 2012.

11. Bradley Willcox, "Caloric Restriction, the Traditional Okinawan Diet, and Healthy Aging," *Annals of the New York Academy of Sciences 1115*, 2007, 434–55.

12. "Caloric Intake from Fast Food Among Adults: United States, 2007–2010," Center for Disease Control NCHS Data Brief, Number 114, February 2013, www.cdc.gov/nchs/data/databriefs/db114.htm.

13. 1900–1970, US Public Health Service, *Vital Statistics of the United States*, annual, Vol. I and Vol II; 1971–2001, US National Center for Health Statistics, *Vital Statistics of the United States*, annual; *National Vital Statistics Report (NVSR)* (formerly *Monthly Vital Statistics Report*); and unpublished data, www.infoplease.com/ipa/A0922292.html.

14. Worldwide HIV & AIDS statistics, December 31, 2012, www.avert.org/worldstats.htm.

15. "Once Aids Was a Death Sentence. Now It's Become a Way of Life," *The*

*Independent*, December 1, 2010, www.independent.co.uk/life-style/health-and-families/health-news/once-aids-was-a-death-sentence-now-its-become-a-way-of-life-2148095.html.

16. NPR, "Two More Nearing AIDS 'Cure' After Bone Marrow Transplants, Doctors Say," July 26, 2012, www.npr.org/blogs/health/2012/07/26/157444649/two-more-nearing-aids-cure-after-bone-marrow-transplants-doctors-say.

17. "Dramatic Increase in Survival Rates for Some Cancer Types, Study Shows," *The Guardian*, November 22, 2011, www.guardian.co.uk/world/2011/nov/22/increase-survival-rates-cancer-types.

18. "A Crowded World's Population Hits 7 Billion," Reuters, October 31, 2011, www.reuters.com/article/2011/10/31/uk-population-baby-india-id USLNE79U04N20111031.

19. "Report: 40 Percent of Newborn Girls Will Live to 100," *U.S. News*, January 23, 2012, www.usnews.com/news/articles/2012/01/23/report-40-percent-of-newborn-girls-will-live-to-100.

20. United Nations, "World Population to 2300," New York, 2004, 1.

21. Ibid., 2.

22. Ibid., 27.

23. *Transcendent Man: The Life and Ideas of Ray Kurzweil*, Ptolemaic Productions, 2009.

24. "Generation of a Synthetic Memory Trace," *Science* 335 (6,075), March 23, 2012, 1,513–16.

25. "Reconstructing Visual Experiences from Brain Activity Evoked by Natural Movies," *Current Biology* 21 (19), September 22, 2011, www.cell.com/current-biology/abstract/S0960-9822%2811%2900937-7.

26. "Dramatic Increase in Survival Rates for Some Cancer Types, Study Shows," *The Guardian*.

27. "Five Things We Learned at Ray Kurzweil's Immortality Lecture," The Grid, October 24, 2012, www.thegridto.com/city/local-news/five-things-we-learned-at-ray-kurzweil%E2%80%99s-immortality-lecture/.

28. David S. Jones, "The Burden of Disease and the Changing Task of Medicine," *New England Journal of Medicine*, June 21, 2012, www.nejm.org/doi/full/10.1056/NEJMp1113569.

29. Author's interview with Eric Topol, June 2012.

30. According to Topol.

31. "IBM's Watson Supercomputer to Diagnose Patients," *Computerworld*, September 12, 2011, www.computerworld.com/s/article/9219937/IBM_s_Watson_supercomputer_to_diagnose_patients.

32. "Exclusive: Time Talks to Google CEO Larry Page About Its New Venture to Extend Human Life," *Time*, September 18, 2013, http://business.time.com/2013/09/18/google-extend-human-life/.

## 4. JOBS: A MILLION LITTLE GOOGLES

1. "Better than Human: Why Robots Will—and Must—Take Our Jobs," *Wired*, December 24, 2012, www.wired.com/gadgetlab/2012/12/ff-robots-will-take-our-jobs/all/.

2. Ibid.

3. CNET, "The Noodlebot Takes on China's Noodle Chefs," August 28, 2012, http://asia.cnet.com/the-noodlebot-takes-on-chinas-noodle-chefs-62218476.htm.

4. Singularity Hub, "Robot Car Wars: Nissan Jumps into the Fray with Driverless Car by 2020," September 9, 2013, http://singularityhub.com/2013/09/09/robot-car-wars-nissan-jumps-into-the-fray-with-driverless-car-by-2020/.

5. Andrew N. Rowan, *The State of the Animals IV: 2007* (Washington, DC: The Humane Society of the United States, 2007), 176.

6. Erik Brynjolfsson and Andrew McAfee, *Race Against the Machine: How the Digital Revolution Is Accelerating Innovation, Driving Productivity, and Irreversibly Transforming Employment and the Economy* (Lexington, MA: Digital Frontier Press, 2011), 28.

7. Ibid., 29.

8. Ibid., 35.

9. "Jobless Rate Edges Down to Its Lowest Level in 4 Years," *New York Times*, December 7, 2012, www.nytimes.com/2012/12/08/business/economy/us-creates-146000-new-jobs-as-unemployment-rate-falls-to-7-7.html?_r=0.

10. Brynjolfsson and McAfee, *Race Against the Machine*, 58.

11. Author's interview with Jonathan Medved, October 2012.

12. Dan Senor and Saul Singer, *Start-Up Nation: The Story of Israel's Economic Miracle* (Toronto: McClelland and Stewart, 2009), 11–13.

13. Ibid., 15.

14. Secondary author interview with Jonathan Medved, February 2013.

15. "Technology Sector Found to Be Growing Faster than Rest of US Economy," *The Guardian*, December 6, 2012, www.guardian.co.uk/business/2012/dec/06/technology-sector-growing-faster-economy.

16. "Start-ups—Present and Future," CIBC report, September 25, 2012.

17. Global Entrepreneurship Monitor 2012 Global Report, 28.

18. Author's interview with Colin Angus, April 2008.

19. "Start-ups—Present and Future," CIBC report.

20. Brynjolfsson and McAfee, *Race Against the Machine*, 58.

21. United Way, "It's More than Poverty: Employment Precarity and Household Well-being," February 2013, www.unitedwaytoronto.com/downloads/whatwedo/reports/ItsMoreThanPovertySummary2013-02-09singles.pdf.

22. "Share of the Work Force in a Union Falls to a 97-Year Low, 11.3%," *New York Times*, January 23, 2013.

## 5. ARTS: LONG LIVE THE DEAD BUFFALO

1. Estimates come from Jonathan Good's 1000 Memories blog, "How Many Photos Have Ever Been Taken?" September 11, 2011, http://blog.1000memories .com/94-number-of-photos-ever-taken-digital-and-analog-in-shoebox.

2. A quick history of digital cameras can be found in my book *Sex, Bombs, and Burgers: How War, Pornography, and Fast Food Have Shaped Modern Technology* (Guilford, CT: Lyons Press, 2011), 179–81.

3. "There Will Soon Be One Smartphone for Every Five People in the World," *Business Insider*, February 7, 2013, www.businessinsider.com/15-billion-smart phones-in-the-world-22013-2.

4. Ibid.

5. Ibid.

6. Facebook blog, "Capturing Growth: Photo Apps and Open Graph," July 17, 2012, https://developers.facebook.com/blog/post/2012/07/17/ capturing-growth--photo-apps-and-open-graph/.

7. The literacy figure comes from "Global Rate of Adult Literacy: 84 Percent, but 775 Million Still Can't Read," *The Globe and Mail*, September 7, 2012, www .theglobeandmail.com/news/world/global-rate-of-adult-literacy-84-percent- but-775-million-people-still-cant-read/article4528932/. The books published figure is from UNESCO, "World Information Report 1997/1998," UNESCO Publishing, 1997, 318.

8. UNESCO, "World Information Report 1997/1998," 320.

9. Floor 64 report, "The Sky Is Rising," January 2012, 2.

10. "Self-Publishing Sees Massive Growth," *The Guardian*, October 25, 2012, www.guardian.co.uk/books/2012/oct/25/self-publishing-publishing.

11. Nielsen, "Buzz in the Blogosphere: Millions More Bloggers and Blog Read-ers," August 3, 2012, www.nielsen.com/us/en/newswire/2012/buzz-in-the- blogosphere-millions-more-bloggers-and-blog-readers.html.

12. "Global Rate of Adult Literacy," *The Globe and Mail*.

13. "Recording History: The History of Recording Technology," retrieved on April 23, 2012, www.recording-history.org/HTML/musicbiz2.php.

14. Brian J. Hracs, "A Creative Industry in Transition: The Rise of Digitally Driven Independent Music Production," *Growth and Change* 43 (3), Wiley Periodicals, September 2012, 444.

15. Ibid., 445.

16. Ibid., 446.

17. "Business Matters: 75,000 Albums Released in U.S. in 2010—Down 22% from 2009," Billboard.com, February 18, 2011, www.billboard.com/biz /articles/news/1179201/business-matters-75000-albums-released-in-us-in -2010-down-22-from-2009.

18. "Massive Growth in Independent Musicians & Singers Over the Past Decade," Techdirt, May 30, 2013, www.techdirt.com/blog/casestudies/articles/ 20130529/15560423243/massive-growth-independent-musicians-singers-over- past-decade.shtml.

19. Hracs, "A Creative Industry in Transition," 453–54.

20. Floor 64, "The Sky Is Rising," 24.

21. Artists House Music, "Did More People Make Music in the Past?" November 2007, www.artistshousemusic.org/videos/did+more+people+make+music+in+the+past.

22. UNESCO Institute for Statistics, "From International Blockbusters to National Hits," February 2012, 8–10.

23. Peter Lev, *The Fifties: Transforming the Screen, 1950–1959* (Berkeley: University of California Press, 2006), 149. The Indian number comes from Cineplot, http://cineplot.com/indian-films-1941-1950/.

24. Scott Kirsner, *Inventing the Movies: Hollywood's Epic Battle Between Innovation and the Status Quo, from Thomas Edison to Steve Jobs* (Boston: CinemaTech Books, 2008), 199.

25. Numbers are drawn from "The Sundance Odds Get Even Longer," *New York Times*, January 16, 2005, http://donswaim.com/nytimes.moviescripts.html, *The Hollywood Reporter*, "Sundance Film Festival Unveils 2012 Competition Lineup," November 30, 2011, www.hollywoodreporter.com/risky-business/sundance-festival-2012-lineup-267587, and Benjamin Craig, *Cannes: A Festival Virgin's Guide* (London: Cinemagine Media Publishing, 1999), 68.

26. YouTube press statistics, accessed on April 16, 2012, www.youtube.com/yt/press/statistics.html.

27. Author's interview with Patrick Boivin, April 2013.

28. Entertainment Software Association facts, www.theesa.com/facts/, accessed on April 30, 2013.

29. Floor 64, "The Sky Is Rising," 31.

30. AtariAge.com lists 418 as the total number of games for the 2600,http://atariage.com/forums/blog/279/entry-5704-how-many-atari-2600-games-are-there-answer-418/.

31. Floor 64, "The Sky Is Rising," 32.

32. Ibid., 1.

33. Media Molecule blog, "LittleBigPlanet: The Road to 7 Million Levels," www.mediamolecule.com/blog/article/littlebigplanet_the_road_to_7_million_levels/.

34. Author's interview with David Smith, March 2013.

## 6. RELATIONSHIPS: SUPERFICIAL DEGREES OF KEVIN BACON

1. OECD family database 2010, www.oecd.org/els/soc/oecdfamilydatabase.htm#structure.

2. Ibid.

3. Barbara Settles, "The One Child Policy and Its Impact on Chinese Families," prepared for XV World Congress of Sociology in Brisbane, 2002, 18.

4. Ibid., 15.

5. Ibid., 9.

6. Ibid., 12.

7. James Lloyd, "The State of Intergenerational Relations Today," London, ILC-UK, 2008, 1–3.

8. Ibid., 4–19.

9. "Social Networks, Small and Smaller," *New York Times*, April 14, 2012, www.nytimes.com/2012/04/15/business/path-familyleaf-and-pair-small-by-design-social-networks.html.

10. Facebook blog, "Anatomy of Facebook," November 21, 2011, www.facebook.com/notes/facebook-data-team/anatomy-of-facebook/10150388519243859.

11. "You Gotta Have Friends? Most Have Just 2 Pals," Reuters, November 4, 2011, http://vitals.nbcnews.com/_news/2011/11/04/8637894-you-gotta-have-friends-most-have-just-2-true-pals, and Live Science, "Americans Lose Touch, Report Fewer Close Friends," June 23, 2006, www.livescience.com/846-americans-lose-touch-report-close-friends.html.

12. QECD family database 2010.

13. NPR, "Marrying Age in the United States," www.npr.org/news/graphics/2009/jun/marriage/.

14. OECD family database 2010.

15. "Tying the Knot: The Changing Face of Marriage in Japan," *Trends in Japan*, July 28, 1998, http://web-japan.org/trends98/honbun/ntj980729.html.

16. "Indians Swear by Arranged Marriages," *India Today*, March 4, 2013, http://indiatoday.intoday.in/story/indians-swear-by-arranged-marriages/1/252496.html.

17. "Finding Harmony Between Marriage and Technology," *Huffington Post*, April 24, 2011, www.huffingtonpost.com/steve-cooper/finding-harmony-between-m_b_841496.html.

18. "Wives Who Confide Their Troubles Are Found to Strengthen Marriages," *New York Times*, July 12, 1966.

19. Online dating numbers found at Statistic Brain, www.statisticbrain.com/online-dating-statistics/.

20. Ibid.

21. "Online Dating Sheds Its Stigma as Losers.com," *New York Times,* June 29, 2003, www.nytimes.com/2003/06/29/us/online-dating-sheds-its-stigma-as-loserscom.html?pagewanted=all&src=pm.

22. Author's interview with Noel Biderman, June 2013.

23. Kinsey Institute, "Frequently Asked Sexuality Questions to the Kinsey Institute," www.kinseyinstitute.org/resources/FAQ.html#relation.

24. Jeremy Greenwood and Nezih Guner, "Social Change: The Sexual Revolution," *International Economic Review* 51 (4), November 2010, 893–915.

25. "Average Man Has 9 Sexual Partners in Lifetime, Women Have 4," *The Telegraph,* December 15, 2011, www.telegraph.co.uk/women/sex/sexual-health-and-advice/8958520/Average-man-has-9-sexual-partners-in-lifetime-women-have-4.html.

26. Author's interview with Bryant Paul, May 2013.

27. "A Cock and Bull Story," *Slate,* September 2, 2006, www.slate.com/articles/
arts/the_undercover_economist/2006/09/a_cockandbull_story.html.
28. "Is Anal the New Oral?" *Marie Claire,* May 3, 2011, www.marieclaire.com/
sex-love/relationship-issues/anal-sex-popularity.
29. Author's interview with Sherry Turkle, August 2012.

## 7. IDENTITY: GOD IS THE MACHINE

1. Author's interview with Tracy Ann Kosa, May 2013.
2. Tracy Ann Kosa, Khalil el-Khalib, and Stephen Marsh, "Measuring Privacy,"
*Journal of Internet Services and Information Security* 1(4), November 2011, 60–73.
3. Author's interview with John Weigelt, June 2012.
4. Author's interview with Jennifer Stoddart, June 2012.
5. "'Sexual Depravity' of Penguins That Antarctic Scientist Dared Not
Reveal," *The Guardian,* June 9, 2012, www.guardian.co.uk/world/2012/jun/09/
sex-depravity-penguins-scott-antarctic.
6. Author's interview with Ian Kerr, June 2012.
7. Pew Research Center, "Teens, Social Media and Privacy," May 21, 2013, 8, 47.
8. Syracuse.com, "Police Cameras Studies Look at Long-Term Effects
on Crime," March 5, 2013, www.syracuse.com/news/index.ssf/2013/03/
police_cameras_studies_look_at.html.
9. Mikael Priks, "Do Surveillance Cameras Affect Unruly Behavior? A Close
Look at Grandstands," CESifo Working Paper Series No. 2289, April 2008,
www.ne.su.se/polopoly_fs/1.40116.1320678202!/menu/standard/file/Paper-
2Cameras.pdf.
10. io9, "People Under Surveillance Are More Likely to Condemn 'Bad Behavior' in Others," June 17, 2011, http://io9.com/5813160/people-under-surveil
lance-are-more-likely-to-condemn-bad-behavior-in-others.
11. Author's interview with Ron Deibert, June 2013.
12. American Civil Liberties Union, "Does Surveillance Affect Us When
We Can't Confirm We're Being Watched? Lessons from Behind the
Iron Curtain," October 15, 2012, www.aclu.org/blog/national-security/
does-surveillance-affect-us-even-when-we-cant-confirm-were-being-watched.
13. Cartome.org, "Panopticon," http://cartome.org/panopticon2.htm.

## 8. BELIEF: ARE ONE-EYED CYLONS MYOPIC?

1. Author's interview with Noh Yu, July 2013.
2. Pippa Norris and Robert Inglehart, "Gods, Guns and Gays," Public Policy
Research, January 2006.
3. Gallup, "Religiosity Highest in World's Poorest Nations," August 31, 2010,
www.gallup.com/poll/142727/religiosity-highest-world-poorest-nations.aspx.
4. CBC.ca, "Do Countries Lose Religion as They Gain Wealth?" March 12, 2013,
www.cbc.ca/news/world/story/2013/03/05/f-religion-economic-growth.html.
5. Tom Smith, "Beliefs About God across Time and Countries," report for ISSP
and GESIS, April 18, 2012.

6. CBC.ca, "Do Countries Lose Religion as They Gain Wealth?"

7. Daniel Abrams and Haley Yaple, "A Mathematical Model of Social Group Competition with Application to the Growth of Religious Non-Affiliation," arXiv:1012.1375v2, January 14, 2011.

8. Author's interview with Anne Rice, June 2013.

9. Author's interview with Shane Schick, April 2013.

10. "Hard Times in the Death Business," *Business Weekly*, May 8, 2012, http://businessweekly.readingeagle.com/hard-times-in-the-death-business/.

11. "Cremations Increase as Attitudes Change," *Worldwide Religious News*, June 14, 2010, http://wwrn.org/.
articles/33616/?&place=north-america&section=other.

12. "Hard Times in the Death Business," *Business Weekly*.

13. James N. Gardner, *The Intelligent Universe: AI, ET, and the Emerging Mind of the Cosmos* (Franklin Lakes, NJ: New Page Books, 2007), 15.

## 9. HAPPINESS: IT'S ALWAYS SUNNY IN COSTA RICA

1. "Addiction: a Loss of Plasticity in the Brain?," *Science Daily*, June 25, 2010, www.sciencedaily.com/releases/2010/06/100624140912.htm.

2. "World Happiness Report," edited by John Helliwell, Richard Layard, and Jeffrey Sachs, The Earth Institute, 2012, 19.

3. Ibid., 70.

4. Ibid., 71.

5. "World Giving Report," Charities Aid Foundation, 2012, 6.

6. The World Bank and CIA both measure it, but differently.

7. Legatum Institute, "The 2012 Legatum Prosperity Index," www.prosperity.com/Subindexes-2.aspx.

8. Marriage rates are from "World Happiness Report," 76, while religion rates are from Nationmaster.com, www.nationmaster.com/country/da-denmark/rel-religion.

9. Author's interview with Peter Gundelach, March 2013.

10. Author's interview with Alexander Kjerulf, March 2013.

11. Author's interview with Tor Nørretranders, March 2013.

12. "World Happiness Report," 111.

13. Human Rights House, "Plight of the Lhotshampas," July 11, 2007, http://humanrightshouse.org/Articles/8133.html.

14. "World Happiness Report," 145.

## 10. CONCLUSION: MARX WAS RIGHT (SORT OF)

1. Author's interview with John Seely Brown.

2. "Why We're Shutting Off Our Comments," *Popular Science*, September 24, 2013, www.popsci.com/science/article/2013-09/why-were-shutting-our-comments.

3. Robert Wright, *Nonzero: The Logic of Human Destiny* (New York: Pantheon Books, 2000), 7.

4. "Adopting Kin Enhances Inclusive Fitness in Asocial Red Squirrels," *Nature Communications*, issue 1, article no. 22, June 1, 2010, www.nature.com/ncomms/journal/v1/n3/full/ncomms1022.html.

5. "Evolution of Adaptive Behaviour in Robots by Means of Darwinian Selection," PLOS.org, January 26, 2010, www.plosbiology.org/article/info%3Adoi%2F10.1371%2Fjournal.pbio.1000292;jsessionid=68EAEA2BB53C385BD6D7A0A1C47860D6.ambra02.

# INDEX

# ABOUT THE AUTHOR

Peter Nowak has been writing about technology for more than a decade. He has been a staff reporter for the Canadian Broadcasting Corporation, the Canadian *National Post*, and the *New Zealand Herald*, and his work has appeared in top newspapers around the world, including the *Boston Globe*, *Toronto Star*, and *Sydney Morning Herald*. He won the Canadian Advanced Technology Alliance Award for excellence in reporting and was named technology journalist of the year by the Telecommunications Users Association of New Zealand. He lives in Toronto.

CREDIT: RYAN VAN EERDEWIJK